Jie Guo

Reshaping Chinese Cities

Geographie
Geography

Band 27

LIT

Jie Guo

Reshaping Chinese Cities

Neoliberal Transition, Embedded Contestation,
and Urban Renewal of Lanzhou

LIT

Cover image: The picture shows the panorama of Lanzhou
(Jie Guo, May 2020)

This study was supported by China Scholarship Council, China's National Key Research and Development Program (2019YFB2103101), GDAS' Project of Science and Technology Development (2020GDASYL-20200103013, 2016GDASRC-0101).

Bibliographic information published by the Deutsche Nationalbibliothek
The Deutsche Nationalbibliothek lists this publication in the Deutsche Nationalbibliografie; detailed bibliographic data are available in the Internet at http://dnb.dnb.de.
ISBN 978-3-643-90949-7 (pb)
ISBN 978-3-643-95949-9 (PDF)
Zugl.: Heidelberg, Ruprecht-Karls-Universität, Diss., 2016

A catalogue record for this book is available from the British Library.

© LIT VERLAG GmbH & Co. KG Wien,
Zweigniederlassung Zürich 2020
Flössergasse 10
CH-8001 Zürich
Tel. +41 (0) 76-632 84 35
E-Mail: zuerich@lit-verlag.ch   http://www.lit-verlag.ch
Distribution:
In the UK: Global Book Marketing, e-mail: mo@centralbooks.com
In North America: Independent Publishers Group, e-mail: orders@ipgbook.com
In Germany: LIT Verlag Fresnostr. 2, D-48159 Münster
Tel. +49 (0) 2 51-620 32 22, Fax +49 (0) 2 51-922 60 99, e-mail: vertrieb@lit-verlag.de

Dedicated to my daughter Dong Xingge and the people who gave me selfless help in the past.

Summary

With the rapidly expanding global monetary system and the omnipresent crossing national/regional investment in the last three decades, neoliberal ideology, has followed the trajectory of capital investment and penetrated to every corner of the world, exerting extensive influences on the socio-spatial structures and urban fabrics everywhere. This has captured the widespread concern of scholars in relevant disciplines, in particular those who have strong interests in cities of the Global South. This book seeks to contribute to the understanding of urban transformation in the context of global and domestic politico-economic changes and to consider how such interactions of global and indigenous ideologies as well as policy configurations have impacts on the urban dimension through a case study of China.

In investigating this central idea, this study has selected the urban renewal and land redevelopment of Lanzhou for investigation. The discussions focus primarily on two aspects, namely the governance transition from a state-centered, managerialist mode to a place-based, entrepreneurial-like one in the context of China's neoliberal steering; and the emerging elite-led, capital-driven urban renewal projects and speculative land development in Lanzhou, a traditional industrial area. First, I seek to show how neoliberal discourses and policies are integrated with China's existing ideologies and discourses, political-institutional arrangements, and the behavior notions of actor constellations. By using an institutional analysis approach, I call China's institutional transition a "neoliberal steering with a selective logic." This means its institutional transition is a common result of both the games between domestic interest groups and the increasingly intense global-local interactions.

Subsequently, I examine the relationship between China's institutional transition and the rise of entrepreneurial-like governance at the urban level, as well as the nexus between land finance and elite-led, capital-driven urban development. I argue that the universal "contradictory duality" which existed in many spheres of China's politico-institutional system is the fundamental genesis of entrepreneurial-like governance, which is simultaneously equipped with entrepreneurial and regulatory characteristics. This thereby brings about government-led capital accumulation relying on a speculative land development.

Further, this study emphasizes the multiplicity of local governance in producing urban space. It refers to an understanding beyond the analytical framework of governance regimes as singular and mutually exclusive. I take a relational approach, and through an examination of the interaction between local authorities and other stakeholders (i.e., relocated enterprises and private developers) in the renewal process of traditional industrial areas, I demonstrate how Lanzhou's municipal government, allied with the central government as well as other administrative units, has mobilized and operationalized its renewal projects and how the resistance and struggles from other stakeholders play a role in disturbing the municipal government's original plans.

In the two specific cases based on fieldwork conducted in 2011 and 2013, I seek to reveal a benefit-distribution mechanism designed deliberately by Lanzhou's municipal government when it was confronted with competitive bargaining of stakeholders. The findings suggest that through this government-led mechanism, all stakeholders have eventually voluntarily or involuntarily gotten involved in the speculative redevelopment processes and reached alliances through a variety of beneficial connections. However, due to the inherent instability of this mechanism, this kind of partnership is rather fragile and easily broken. Hence, I argue that under the current governance regimes, competition, negotiation and cooperation are an eternal theme. Reflected in the ongoing restructuring of interest relationships, this affects the manners and consequences of the production of urban space thereafter.

Contents

List of Figures

List of Tables

The abbreviation

CAS	Cadre Appraisal System
CLLL	CNR Lanzhou Locomotive Ltd.
CNCs	CPC's National Congresses
CNPS	China National Petroleum Corporation
CPC	Communist Party of China
FSID	The Fifth State-level Industrial District
GPMI	Gansu Province Material & Industrial Group
HCIS	Housing Credit Insurance System
HMP	Housing Monetization Policy
HPFS	Housing Provident Funds System
LIDCC Company	Lanzhou Industrial Development & Construction
LITF	Lanzhou Investment & Trade Fair
LLG	Lanzhou LS Group Co., Ltd.
LLRB	Lanzhou Land Resources Bureau
LLRIC	Lanzhou Land Reserve & Investment Centre
LNID	Lanzhou New Industrial District
LPC	Lanzhou Petrochemical Company
LRPL	Lanzhou Run'an Properties Ltd.
LTBs	Local Taxation Bureaus
LTCL	Lanzhou Textile Co., Ltd.

LUDIC Corporation	Lanzhou Urban Development & Investment
LXDC	Lanzhou Xiaotian'e Development Company
SASAC Company	State-Owned Assets Supervision and Administration
SAT	State Administration of Taxation
SFPs	State-owned Financing Platforms
SLCDC Company	SASAC Lanzhou Construction & Development
TSS	Tax Sharing System

Chapter 1 Introduction

1.1 Research background

Urbanization in the neoliberal era

Since the world ushered in the era of neoliberalism, the manner of global economic growth has shifted from production-sector driven to financial-sector driven (Smith, 2002, p430). Neoliberal ideology is committed to an unfettered capital market, following immediately the capital trajectory that has been penetrating into every corner of the world. Recalling the past thirty years in many cities and regions—although it has to be admitted that this ideology exists in numerous variations in different places—apparently, the core value of neoliberalism and its paradigm of practices in policy-making and strategic planning, have indeed influenced the way of space development and reshaped the geographical landscapes in many regions and areas.

Cities have become important laboratories for a variety of neoliberal policies (Brenner and Theodore, 2002b, p368; Peck, Theodore and Brenner, 2013, p1096), and almost all policy experiments are initiated to "generate future growth, and to wage a competitive struggle to attract capital investment in the intensified global competition" (Swyngedouw et al., 2002, p546). A notable feature of the policy-making process in neoliberal cities is "less democratic and more elite-driven" (Swyngedouw et al., 2002, p542). In their strategies, urban space is regarded increasingly as a "privatized space" that is more and more consistent with the elite's taste. The overarching goal is to mobilize space as an arena both for capital-led economic growth and for the practices of commercial consumption (Brenner and Theodore, 2002, p368).

Thanks to the restless capital flows and the increasingly enhanced nexus between global and local, as Smith (2002, p431) puts it, we are now "experiencing the emergence of a new urbanism," through which "the containers themselves are being fundamentally recast." One of the most profound changes that cities have been experiencing in recent decades and at present is that both international and domestic capital is flooding into the urban built environment at an unprecedented scale and rate and is being fixed in space in some form. By virtue of its scarcity, urban land and real estate are naturally favored by, and also subordinated to the capital.

Stimulating the appreciation of, or at least retaining the value of, urban land and property through a steady flow of capital investment becomes a powerful means of wealth creation. Capital circulates through the built environment in a dynamic and erratic fashion (Weber, 2002, p520), and cities are gradually evolving as "a volatile territory of capital crisis" (Lin, 2014, p1816). It is therefore clear that the urbanization process has been intrinsically connected to "the logic and dynamics of capital accumulation" (Lin, 2013, p50) and turned into a "profitable business" grounded upon the resources of the urban environment (Lin, 2014, p1815). One of the most obvious direct consequences of such capital-led, profit-driven urbanization is the hungrily speculative development of urban resources, particularly the capitalized land development and hyper-commercialization of urban space. Cities are like standardized "production workshops" in which numerous luxury residential areas, five-star hotels, and skyscrapers are assembled in the pipeline.

Against this background, many cities in the world, are experiencing unprecedented uneven development, gentrification, and restless transformation (Brenner and Theodore, 2002; Leitner et al, 2007; Shin, 2015). Rapid urban change comes via mega-reconstruction projects. As we have seen, countless financial centers, shopping malls, tourist attractions, private luxury accommodations, and industrial concentration zones are springing up in many regions and cities, gathering in the city center at the beginning and sprawling later into the peripheral areas. At almost the same time, large amounts of degraded residential neighborhoods, old industrial areas, and landscapes whose functions or appearances are deemed outdated are being removed and eliminated. The reconstruction projects, undertaken either with the purpose of commercial and high-end residential development or for the aim of industrial upgrading, are all striving to cater to the needs of the times. Extracting new value from old urban environments seems to be one of the most fundamental objectives of urban renewal.

As such, contemporary urbanization should be best attributed to the contradictory, hierarchical social relations in the institutional forms of capitalism, through which space continues to be shaped and contested (Brenner, Madden, Wachsmuth, 2011, p225). In urban areas, there exist different interest groups. Each of them holds its own rights and powers of governance. Simultaneously, in the process of commodification of urban space and the land and property development projects, they also have distinct claims and expectations on the development methods, uses of space and distribution of benefits. Thus, accompanied by the constant upgrades of landscape and enormous changes of urban function, the balances of social forces and power relations are undergoing tremendous changes. This process is full of cruel deprivation and occupation, fierce resistance and struggle, and inevitable compromise and/or proactive cooperation, in which some of the social forces are deprived of interests, while others benefit from it. With the continuously intensifying and relentless clashes, social forces start to be layered. Some of them rise to dominance, while some others are relegated to a subordinate position or even at risk of being marginalized (Brenner, Marcuse and Mayer, 2009, p178).

Within this trend, new bargaining mechanisms have emerged. Yet they are filled with ongoing contradictions, interim compromises, and unstable cooperation. On the one hand, neoliberalism, as the predominant set of norms, ideas, policies and practices seems to leave limited room for bargaining to social forces beyond the capital market (Brenner and Theodore, 2002b). It thereby leads to the fact that state authority, the capital market and broad-based local forces are extremely unbalanced. On the other hand, the pervasive penetrations of capital into the economic, social, political, and cultural spheres is mobilizing broader interest group, and stimulating their enthusiasm to participate in the feast of wealth creation and profit-sharing. Interests are split into fragments and are shared by distinct stakeholders through intense competitions. Because of it, they are related to each other much more closely than ever before, owing to various benefit associations. Thus, with the growing surge of interactions between politico-economic activities and the urban environment, social relations and power structures are shifting imperceptibly and have become increasingly complicated and diverse. Meanwhile, urbanization as well as urban transformation is moving ahead, driven by a group complex nexus of social forces and multi-faceted power relations. And coordination and integration of multifarious—and sometimes even contradictory—interests and demands on urban space are proving to be the difficulty as well as the focus of local governance (Pierre and Peters, 2006).

Exploring the nature of contemporary urbanism—the need for a critical perspective

The newly emerging and controversial phenomenon stated above demonstrates a global trend of urbanization and urban restructuring in the contemporary geo-economic context. Many critics of the neoliberal model of urban development have argued that it is necessary to take a closer look at how different types of cities are being repositioned across the world system and are undergoing drastic transformation within increasingly volatile circuits of capital accumulation (Brenner, Marcuse and Mayer, 2009, p176). Meanwhile, it appears increasingly urgent to point out the limits of such profit-based urbanism characterized by what Harvey termed "accumulation by dispossession" (Harvey, 1989; Smith, 1996; Brenner and Theodore, 2002b).

In light of these challenges, more and more critical scholars suggest that intellectuals, activists and political elites should carefully re-examine and also forecast this neoliberal urbanization from a critical perspective. At the same time, a particular attention is proposed to be paid to the "post-crisis urban governance," i.e. the critical investigation of "policy-in-motion" across multiple sites (Peck, Theodore and Brenner, 2013, p1096). Moreover, it is imperative is to propose a "revolutionary" urban theory, and to "chart the path" towards an alternative, post-capitalist form of urbanization (Brenner, Marcuse and Mayer, 2009, p177). To this end, it is proposed to open up a more inclusive arena to challenge the mainstream theories and monotonous approaches to urban questions, on which "wide-ranging disagreements about any number of core theoretical, methodological and political issues" are possible to be discussed (Brenner, Marcuse and Mayer, 2009, p179).

Generally speaking, establishing a framework for critical analysis is suggested to start from but not be limited to the following points. First, "place-based investigations" are highly suggested (Peck, Theodore and Brenner, 2013, p1096), namely those linking global processes of socio-economic restructuring with local contexts. Observing diverse evolutionary paths of cities in the common global politico-economic context is an effective way to identify the intrinsic nature of urban restructuring. It suggests that we must pay attention to how power relations and regulatory ideologies, practices, and institution in specific places condition the evolution of urban regimes and urban transformation. Undoubtedly, neoliberal discourse is still hegemonic in the global, but this does not mean that the dominant ideological paradigm reigns uncontested and is unchallenged (Mayer and Künkel, 2012, p3; Peck, Theodore and Brenner, 2013, p1093). On the contrary, when it is locally fixed and has to compete with indigenous ideologies and inherent institutional arrangements and challenge the existing power- and interest-based relationships, thereby evolving into new variants that are different from its universal policy paradigm. Thus, a study of historical-cultural contexts, institutional arrangements and socio-economic conditions in specific places is necessary. It provides a way to find the local origins of variations of the global paradigm, which leads urbanization and urban restructuring in different directions. Only in this way, is it possible to give reasonable explanations for various evolutionary paths of urbanization within a same global framework of capitalism (Peck,

Theodore and Brenner, 2013, p1096; Brenner, Marcuse and Mayer, 2009, p178-179).

Second, the study of the urbanization process and urban transformation in the age of neoliberalism must examine the pattern of governance within a systemic socio-political context—i.e. an extra-local understanding of governance, and simultaneously in a particular historical period (Brenner, Marcuse and Mayer, 2009, p179; Peck, Theodore and Brenner, 2013, p1096). In particular, an exhaustive and detailed investigation of specific practices of local governance and the strategies of local elites with a selective logic, in face of both global politico-economic changes and local realities, is necessary. Governance is an intermediary that connects global and local discourses, and actors within the framework of governance are responsible for the transmission of information. Some regulatory practices and policy apparatus of governance are highly praised as a universal model (e.g. local entrepreneurism) and then are selectively used by actors in other places as references for discovering and solving local problems. In this way, knowledge about urban problems and the ideas and technologies of addressing them are transferred from one place to another (Peck, Theodore and Brenner, 2013, p1096; McCann and Ward, 2011). Thus, by tracking the "policy-in-motion" across multiple sites (Peck, Theodore and Brenner, 2013, p1096), it is possible to capture more detailed and vivid process of how global discourses and policy configurations mutate in different cities/regions. Accordingly, in empirical studies, two questions are interesting and worth exploring—that is, "how globally circulating discourses and policy models are fed into the socio-economic configuration and political actions of local and how, conversely, different urban

7

contexts shape the adoption of global discourses and policy models at the local level" (McCann and Ward, 2011).

Third, the critical analysis is suggested to pay close attention to the "ongoing and emerging socio-political struggles" all the time (Brenner, 2009, p200). As stated above, the process of neoliberalization is not unchallengeable. Instead, "oppositional and challenging social movements have everywhere impacted on this multifaceted process of generating locally specific forms of neoliberal urbanism, very often oscillating ambiguously between resistance and integration" (Mayer and Künkel, 2012, p3). Therefore, it requires a systematic research framework from a new perspective, which involves a comprehensive understanding of 1) the macro-structural and macro-regulatory forces that shape the evolving "rules of the game" (Peck, Theodore and Brenner, 2013, 1097); 2) the microphysics of governance that implement and operate these rules at the local level through particular projects and strategies, as well as innovative techniques (Uitermark, 2005, p142); and 3) the micro-practices of resistance and cooperation in local settings, namely the interactions between local stakeholders and the possible and subtle impacts on central and local decision-making. Besides, the critical study is asked to look at the "marginalizations, exclusions and injustices that are inscribed and naturalized within existing urban configurations"— i.e. problematizing the sites and sources of social injustice—, and place the pursuit of "an alternative form of urbanism based on the remaining potentials of contemporary cities" as the ultimate goal (Brenner, 2009, p204; Brenner, Marcuse and Mayer, 2009, p 179; Peck, Theodore and Brenner, 2013, p1097).

China as a typical and unique case

China's urbanization processes in recent decades seem to be a suitable object of review, and a rare debate arena for critical urban theory to show its mettle. The phenomenal growth and transformation of Chinese cities, as many scholars (Harvey, 2005; He and Wu, 2009; Wu, 2008, 2016; Lin, 2014; Peck and Zhang, 2013) have argued, is difficult to understand if completely adopting the perspectives of traditional, mainstream theories without any critical reflections on and innovations to these theories. For instance, either the explanation of economic agglomeration and technological innovation in promoting urbanization in traditional urban economy theory (Florida et al. 2012, p645) or the binary opposition of state-market relations in neoliberal political economy theory, turns out to be problematic when these concepts are used directly to analyze China's complex economic and urban phenomenon (Lin, 2014, p1817; Lin and Zhang, 2015, p2775; Peck and Zhang, 2013; Florida, et al. 2012; Wu, 2016). Nevertheless, these theories and concepts have given appropriate explanations for the process of urbanization as well as urban issues in the Global North (Lin, 2014, p1817; Robinson and Roy, 2015; Wu, 2016; Roy and Ong 2011).

In light of critical urban theory, research on Chinese cities is necessary and also meaningful (Robinson, 2011; Wu, 2016; Shin, 2015), since the Chinese paradigm has both typical characteristics of global urbanization and its own uniqueness. On the one hand, when Western countries transformed from Fordism to the post-Fordism economy, along with the politico-economic system steering overwhelmingly from Keynesianism to neoliberalism, China also experienced a drastic transformation from a socialist and pre-industrial society to a more open, market-oriented economy and a modern industrial state. Along with this process, it is the rapid industrial agglomeration, massive rural-urban migration, swift and drastic increases in the amount number and size of cities, social transformation and stratification, and environmental issues as well as other urban diseases. It seems that the country's 50-year history of urban development is more like a microcosm of the various urbanization processes seen around the world in the past 200 years. The urban sprawl and low-density suburbanization that occurred in North America, the urban-rural dual structure of the Soviet Union, the slum cluster and urban poverty of Latin America, and the explosive growth of core cities and speculative renewal that happened in the newly industrialized countries in East Asia, are all easily associated with some corresponding cases in China.

On the other hand, China's case is unique. Its urbanization process displays some features of the market-led development of the Western neoliberal state while simultaneously showing typical features of the developmental states in East Asia (Wu, 2016). Market-driven and state-let paradigms are theoretically far apart, but in reality they are fused together in China's practices. However, contradictions are pervasive in, and obvious in the politico-institutional arrangements, policy-making, and development and governance practices underlying China's ambitious urban transformation. Precisely because of the paradoxical coexistence, Chinese urbanization displays great messiness and uniqueness.

In the politico-institutional sphere, the first contradiction concerns the relationship between the state and the market, or rather the role of the state in social and economic process (Harvey, 2005; He and

Wu, 2009; Lin 2014; Wu, 2008; Zhang, 2012). Since 1978, China has partially opened the market and allowed the private sector to get involved in urban development. However, the state still plays a major role in affecting the speed and scale of urbanization, and promoting urban and social transformations. Secondly, the paradox exists between the devolution of responsibilities to the urban level and the fiscal recentralization of the central state (Wu, 2016; Lin, 2014, 2015; Fu, 2015). The third contradiction is the enhanced power of local governments in economic decision-making and urban affairs, and the remaining intervention of the central state by appointing local officials and evaluating their performances (Lin, et al. 2013; Wu, 2016). Likewise, there are some other paradoxical factors that coexist and interact with each other and jointly account for the complicated, changeable process of urban transformation and the unique socio-spatial phenomenon in China's urbanization. For instance, contradictions exist between the ongoing growth of jobs and the rigid household registration system; between the rising voices and demands of the free market (especially the housing and land market) and the imperfect system of property rights; and between the rapid population growth in cities and the inadequacy of urban housing and infrastructure and welfare provision.

The broad contradictions in politico-institutional arrangements and regulatory practices have brought about several far-reaching consequences in socio-spatial aspects. The most significant phenomenon is the massive rural-urban migration movement, along with the accelerated growth of urban populations[1] and a

---

1   Urbanization rate has risen from 17.9% in 1978 up to 53.7% in 2013, source from: New national urban planning (2014-2020).

severe shortage of urban land resources promoted by the planning and policies of the central state over the last four decades. For instance, in China's latest National Urbanization Plan (2014-2020), the central government introduces a series of incentives related to household registration (hukou), employment, and medical and other welfare services for rural immigrants. Meanwhile, the plan develops national-level strategies for new town construction and urban renewal to expand urban capacity. Its ultimate goal is to facilitate further rural-urban migration involving more than 100 million rural inhabitants.

Another equally striking phenomenon is an enormous expansion of urban construction lands at the expense of encroaching upon agricultural land, along with the recent relentless reconstruction and speculative renewal in both city center and urban fringes. According to official Chinese statistics (National Land-use Planning, 2006-2020), between 1996 and 2011, the total cultivated land decreased substantially from 130.06 to 121.67 million hectares. In contrast, urban built-up areas as well as their "eroded boundaries" extended continuously and rapidly, while eating away at the surrounding rural areas. In recent years, the project to redevelop dilapidated, low-density urban areas took over urban expansion has become the highlight of China's urbanization stage. In two official documents published by the National Council, entitled Guidance on Accelerating Shantytowns Reconstruction of State-owned Enterprises and Notice on Further Strengthening the Shantytowns Reconstruction, a total of 20.8 million units of urban housing and around 15.65 million units of rural housing in shanty towns and urban villages were demolished and reconstructed between 2008 and 2014. Since 2015, the massive demolitions and reconstructions continue to accelerate. These projects involve 213,700 hectares of construction land in the city centers and urban fringes[2]. Simultaneously, the relocation and reconstruction of old industrial areas in around 120 cities are in full swing, which is estimated to free up 1.4 million km[2] of construction lands for these cities[3].

Accompanying the politico-institutional transition, socio-demographic changes, and the ongoing changing of geographical landscape, is the dramatic transformation of local governments as well as their governance practices. Some scholars, by distinguishing Chinese characteristics of neoliberalism, in which the state-market relationship is the focus of attention, have pointed out that the decentralization of power and the marketization of the economy, along with the resulting self-adaptive adjustment of local states should account for the "outstanding" economic performance (Harvey, 2005; Wu, 2008, 2010; Zhang, 2013), and such massive city (re)construction movements (He and Wu, 2009; Hsing, 2010; Wu, 2016;). Lin (2013 and 2014) analyzes the inherent associations of state power reshuffling, urban land development, and local public finance and proposes "a tripolar political economy" to explain the origins of the "first-growth" strategy and the profit-driven behavior of governments at the urban scale. In the literature, various conceptions have been adopted to illustrate such rising localism, such as the local state corporatism (Oi, 1992), local entrepreneurialism (Duckett, 2001; Walder, 1995; Su, 2015), a de

---

2   Data from http://paper.people.com.cn/rmrb/html/2013-07/24/nw.D110000renmrb_20130724_1-02.htm

3   ibid.

facto economic federalism (Qian and Weingast 1997), territorialization of land management (Hsing 2006), and so on.

However, regardless of how different the explanations of these concepts are, one fact is always present. That is, the growth of Chinese cities relies increasingly on the mechanism of the state land monopoly (Wu, 2016) and is driven and accelerated by locally financed and land-based projects (Lin, 2014, p1818). He et al. (2007) and Shi (2015), by studying the process of urban renewal and gentrification, and the emergence of new urban spaces as well as the related social-economic impacts, point out that the commercialization of urban space, the dispossession of land property, and speculative land development are the main features of China's new urbanism. In promoting the renewal process, the real estate sector along with private capital plays a crucial role. Their forces are considered not only as the main engine of the growth of local economy as well as fiscal revenue (Fu, 2015), but also as a dominant instrument to assist local government in initiating mega renewal projects (He and Wu, 2007; Shin, 2015; Wu, 2016) and land-centered urbanization processes (Lin, 2014).

Large-scale urban renewal, along with the influx of massive capital into land, has been proven to be in favor of re-imaging the city and promoting urban transformation (He and Wu, 2007). Meanwhile, it has been widely recognized that the drastic transformation of an urban environment inevitably gives rise to the reshaping of social relations among stakeholders (Wu, 1997; He and Chen, 2012). Different interest groups are interconnected by competing conceptions of what and how places should be produced, and how space should be used (He and Chen, 2012; He and Lin, 2015). Their relationships are strengthened or weakened through various interest chains which are formed on the basis of land and housing property, hierarchical authority, legality in resource allocation, planning and local management, and so on.

However, in regard to the participation of distinct interest groups with their varied power relations, the process of urban renewal and land development as well as the internal mechanism (involving negotiations and distribution of benefits) which promotes this process are quite different. For instance, as many scholars have argued, the process of a residential area's renewal is always accompanied by brutal plunder of wealth and land, social injustice, marginalization of vulnerable groups, and non-formal resistance. However, will these similarly occur in the process of an industrial area's redevelopment? If not, do any other phenomena or serious problems emerge, and what causes and mechanisms lead to or generate these phenomena or serious problems?

1.2 Research objectives and key questions

The objectives of this research are to find out the intersections between institutional transition, urban governance, and urban transformation in China, particularly as these are concerned with 1) the intersections between China's neoliberal transition with a selective logic and elite-led, capital-driven urbanization processes, and 2) the dynamic balance of power relations, public-private interactions, and political-institutional arrangements that are reflected in, and are conversely shaped by, the changes to the urban built environment.

Specifically, this study aims to analyze the collision of various ideologies in contemporary China—i.e. the inherent political-institutional and cultural context and the global neoliberal hegemony—and to find out the genesis of China's neoliberal steering. At the urban level, this study hopes to identify how the hybrid institutional system (exhibiting both neoliberal and authoritarian characteristics) affect urban strategies and policy-making, thereby leading to speculative and frequent urban renewal and land development. It is especially intended to present a kind of new urban governance mode (covering strategic objectives, policy setting and implementation tools) which has emerged recently in response to the state's neoliberal shift. The third objective is to sketch out the process of mobilization of and resistance to urban renewal. Its significance lies in revealing the interaction mechanism hidden behind the frequent conflicts, struggles, negotiations, and cooperation among multiple stakeholders during land redevelopment.

Thus, in the empirical parts of this thesis, discussions focus closely on the following questions:

1) Why are globally neoliberal discourses and general policy paradigms selectively applied in China's system reform, and especially how are they integrated with the existing ideologies and discourses, political-institutional arrangements and behavior notions of actor constellations?

2) What impacts does this hybrid institutional system with both neoliberal and authoritarian features have on China's urban governance? In particular, what objectives and strategies have been proposed by local governments in response to the macro institutional changes? And how are these connected to the elite-led, capital-driven urban development of China?

3) In compliance with municipal blueprints, how do local political elites mobilize a broad participation of society? What kind of technologies and art of governance are adopted, and conversely, what forms of resistance have local authorities encountered?

4) In response to contradictions and conflicts of interest, how has an implicit coordinating mechanism been designed, through which the vast majority of stakeholders have been voluntarily or involuntarily involved in the speculative urban renewal process?

1.3 Research methods

As explained in Section 1.2, this thesis aims to discuss the new form of urban governance in contemporary China, and the process of large-scale urban renewal, land development, and the production of urban space within the framework of this governance regime. All these changes and processes happen within an institutional environment which can be seen as the driving force for any changes in governance and the conditions for stakeholders' participation in urban renewal. Thus, it is necessary to conduct the study of urban transformation in a broader framework, by considering the particular local settings that cities have been rooted in, the global political-economic context, and the inter-organizational factors and complex interactions among various urban subjects.

The system analysis method offers just such a perspective for describing complex environmental situations and the interdependent activities within the environment (Anderberg, 2005, p80). Systems thinking regards the process as a structured whole and emphasizes thinking in a cyclical rather than a linear cause and effect scheme (Heurkens, 2012, p116). It focuses on discussing the relationships, interactions and mechanisms hidden behind dynamic and complex surfaces (Checkland, 1981; see Anderberg, 2005). From the perspective of a systems approach, any group of practices and a set of organizations within a process function as the component parts of a system. These component parts (i.e. the subsystems) are interconnected by certain material, energy or information flows, and are interacted by various traffic exchanges. Similarly, changes in some components of a system will cause changes in other parts. It is also possible to trigger changes in other systems, including a parent system and larger systems. Thus, a system should "be best understood in the context of relationships with each other, and with other systems" (Heurkens, 2012, p16). Further, special attention should be paid to the "changes and transformation and the regulation (control) of the system" (Anderberg, 2005, p81). Inspired by systems thinking, this study outlines a conceptual framework (Figure 1.1). In this framework, cities are considered to be situated in a huge system constituted by the global politico-economic environment, domestic ideologies and institutional arrangements, and interactions in the actor constellation involving multiple scales. Each of them functions as a subsystem constituted by a variety of relational structures, and conveys information to each other. In the complex and huge system, cities constantly receive information flows from both global and national levels (e.g. strategic concepts and political-institutional arrangements), and conversely transfer feedback upward.

Figure 1. 1 Conceptual framework and research objectives

In order to track the locus of information transmission, primarily in relation to the concepts transferring from the global level to the urban level, and to identify the interactions between each section during the transmission, this thesis has employed a number of different approaches (Figure1.2). First, an institutional analysis approach is adopted to discuss how global neoliberal discourses and concepts are fixed at the local level, compete with local ideologies, and then contribute to national institutional transition. Second, from the perspective of a neo-Marxist political economy, this study has examined the relationship between China's institutional transition and the rise of entrepreneurial-like governance at the urban level, and the relationship between land finance and capital-driven urban development. However, as Jessop and Sum (2001, p97) argued, political economy has an impoverished notion of "how subjects and subjectivities are formed and how different modes of calculation emerge and become institutionalized." Thus, this study employs the third approach for analysis, Foucault's governmentality approach. It is used as a supplement to identify the microphysics of this process— i.e. what strategies and techniques are used by national and local governments in compliance with their political blueprints. Furthermore, a relational approach is used to examine the mobilization of local governments and the resistance of relevant stakeholders. It aims to prove a non-binary opposition stance among stakeholders in the urban renewal process.

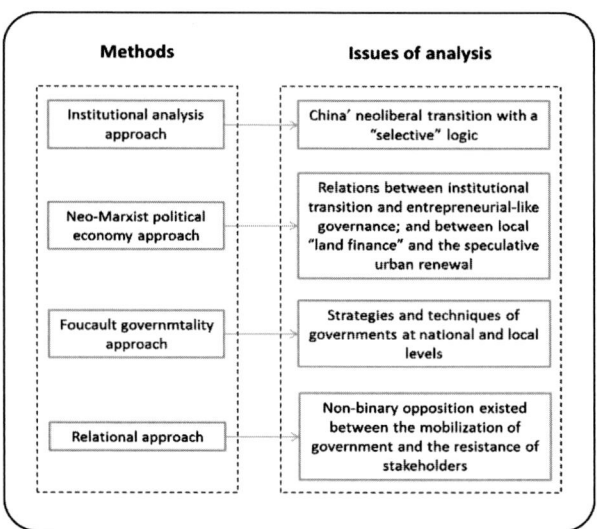

Figure 1. 2 Methodological framework

Research phases and data collection

Around the core issues above, the research is undertaken in three steps. At each stage, relevant documents involving a wide range of textual and visual data were obtained by different techniques. This thesis constructs the theoretical framework by collecting and reviewing academic and professional literatures. Then, it extracts and summarizes the core points of related theories, and defines the research problems. Based on this, this thesis attempts to develop a coherent conceptual framework, by integrating the independent knowledge in relation to the institutional setting, governance mode and production of urban space into a whole object of the study.

In the second step, this study conducts extensive literature and archives reviews on China's institutional transition, particularly referring to the land, housing and fiscal system reforms over the past 40 years. Historical archives are highly valued here, since historical archives are the "thought fragments from the past", which is conductive to uncover the real world overshadowed under the gorgeous veils (Arendt, 1968, p51-52). Specifically, data collection in this step consists of official documents of the previous CPC's National Congresses, and policies as well as statistical data on land, housing and financial reforms issued by the State Council. These documents show how state-led interventions in political-economic spheres had direct and long-lasting effects on land-driven, elite-led urbanization, and local governance transition. Besides, the data used in this part also contains those published by Lanzhou municipal government. It comprises of the programs, policies and data related to urban renewal, and the historical archives about Lanzhou city and economic development (e.g. Chronicles and Statistical Yearbooks).

However, it should be admitted, over-reliance on official documents will lead to biased results. Since in line with Foucault's opinion (1977), any discourse is subject to the domination of a specific power. As such, it is extremely necessary to take multiple discourses into account, rather than a "single word". Thus, as a crucial supplement, the unofficial data and documents were obtained from multiple sources. The unofficial discourse relates to the media reports (e.g. local newspapers, and video interviews published online), and the information from the semi-structured interviews with stakeholders (e.g. managers of relocated enterprises and property developers, serving and retired employees, and local residents). Besides, other textual and visual materials, such as the data and planning schemes released by the unofficial stakeholders—private planning firms, and the photographs made by the author, were collected during the field work between September and November 2013. The way of connecting and incorporating official and unofficial discourses into an overall analytical framework, helps to distinguish, and also to compare the opposing positions and ideologies of different interest groups over some concrete issues—such as what/how places should be produced, what/how it should be used, who enjoy the absolute/part ownerships of space, and who own the rights to distribute and/or occupy the profits generated on space?

1.4 Structure of the book

To answer the questions stated in Chapter 1.2, the remainder of the thesis is organized as follows. Chapter 2 provides the main conceptual and theoretical basis for this thesis. It seeks to build a study framework by examining the intersection between the macro-institutional contexts and urban governance. By reviewing relevant research paradigms in relation to neoliberal urbanism in a global context, this chapter draws on some critical perspectives in discussing the rising entrepreneurialism, and the excessive capitalization of land development, and commercialization of urban space.

Chapter 3 provides an essential connection between the case study and China's institutional context that the study is embedded in. This chapter starts with presenting the fiercely ideological debates over China's neoliberal restructuring in contemporary. Then, the relation between the state and market, and the nexus between the central and local states are examined. In particular, an in-depth analysis is conducted, with respect to the reforms on urban land and housing markets, fiscal system, and official appraisal system. All of these are considered as the most fundamental impetus for China's land-dominated urbanization and "growth-first" governance strategy in the past 40 years. At last, the chapter presents how the new land management system operates under the condition of local fiscal difficulties.

Chapter 4 refers to the case study of Lanzhou. It first gives a brief introduction to the course of industrial development and urban transformation in contemporary Lanzhou, which provides an overview of the historical, geographical and social context for the subsequent discussions. Then, it answers how urban renewal has been launched by Lanzhou municipal government as mega projects with assistance of central and provincial governments; and what kind of mobilization techniques have been employed to attract broad participation of social capital?

Chapter 5 investigates the redevelopment process of industrial areas in Lanzhou, and in which a detailed analysis of two relocation and redevelopment cases is conducted. By revealing the complex interactions among stakeholders, i.e. except for conflicts and struggles but also negotiations and cooperation, this chapter focuses on uncovering a kind of interest-distribution mechanism initiated by the municipal government, through which the vast majority of stakeholders are involved in municipal renewal projects. Finally, Chapter 6 summarizes the findings, offers conclusions and sparks future directions.

Chapter 2 Literature Review and Theoretical Frame: Institutional Transition, Neoliberalism and New Urban Governance

2.1 Institutional reform and new urban governance

2.1.1 Institutional change theory: "new institutionalism" versus "old institutionalism"

As a Chinese saying goes, nothing can be accomplished without norms or standards ("无规矩不成方圆"). It implies that if there are no constraints and rules, human behaviors will fall into chaos. Why do institutions, rules or regulations matter in our social daily life? How do they form and work in social and economic operations as well as physical environment? And what are their real connotations?

The term of institution has been used widespread in the social sciences in recent years, including the philosophy, institutional economics, sociology, political science, geography and other disciplines. There is a growing consensus among scholars from related disciplines. As the persistent and identified structures and the mechanisms of social order, which guide the behavior of a set of individuals or organizations within a given community, institutions do matter to understand important social outcomes (North, 1990; Hodgson, 2006; Walliser, 2006; Etzold, el at., 2012). Obviously, institutions are essential (North, 1990, 1991; Bell, 2002). They perform as a kind of "structures that matter most in the social realm, and make up the stuff of social life" (Hodgson, 2006, p2).

In fact, the term has a long history being studied in the social sciences, "dating back at least to Giambattista Vico in his Scienza

Nuova of 1725" (Hodgson, 2006, p1). However, even today, the debates over the process of institutional formulation and the way of our understanding this process have never stopped. From the perspective of "old institutionalism" (before 1950s), the explanation of institution is more focused on exploring the "formal factors" within it, such as legal, administrative and historical factors and political structures, and the role of these factors played. Formal institutions, were considered to be the formal-legal and administrative arrangements of government and the public sector defined by some form of constitution (Bell, 2002, p4). This interpretation of institution emphasized on the nature of invariance and stability, but "left no room for the analysis of institutional change" (Gross, 1999, p34).

From the late 19th century to the first-half of the 20th century, the analysis of the institutions was more rooted from the perspective of behaviorism. The "behavioral" approaches to politics, in many ways, moved the "research agenda to the alternative pole" (Gross, 1999, p34). In this framework, institutions were "understood as a special type of social structure with the potential to change agents, including changes to their purposes or preferences" (Hodgson, 2006, p2). The related scholars were more convinced that institutions were rooted in human thinking and habits, while the thinking and habits were derived from human instinct. Veblen in his book "Theory of the Leisure Class: An Economic Study of Institutions" published in 1899 (p109) proposed that institutions perform not as organizations, but as "prevalent habits of thought with respect to particular relations and particular functions of the individual and of the community". He considered that the emergence of institutions should be fundamentally attributed to the human instinct (or human nature) (Thorstein Veblen,;). Consistent with the biological evolution theory, the institutions were the results of competition and fittest, which always evolved by adapting to changing environments, but would not be fundamentally changed (Veblen, 1899, p188).

Commons (1934, p87)—another important representative of this period, likewise, adopting the behaviorist approach in his institutional analysis—argued that institutions were, in fact, made up of collective actions that defined the economy, along with conflict of interests. In his book *"Institutional Economics"* (1934), institutions were referred to as "collective actions that were in greater or less control, liberation and expansion of individual action" (1934, p70), and that "ranged all the way from unorganized custom to the many organized going concerns, such as the family, the corporation, the trade association, the trade union, the reserve system, the state" (1931, p649). He inherited Veblen's evolutionary opinion about institution, and also pointed out that the evolution was based on an artificial selection—i.e. designed by collectives for certain purposes, rather than a "natural choice" (ibid).

The "behavior approach" represented by Veblen and Commons stressed excessively on the role of human reason and attitudes, personalities and actions (Gross, 1999, p34) in determining individual's habits, collective customs as well as the whole social institutions. They more inclined to introduce psychological and behavioral factors in the interpretations for institutions. Furthermore, they represented institutions as a set of widely accepted social habits. However, owing to the over-emphasis on the behavior itself, the Behavioral Scholars seemingly ignored the role of contextual attributes (Hodgson, 2006), in which behavior and institutions embedded in—i.e. the changing socio-economic environment, cultural and historical setting, as well as a policy framework. Obviously, actors behave indeed within specific contexts. However, it seems to be unrealistic to give a full and exhaustive definition and explanation on the natural and socio-economic circumstances where they lived in, by only considering the perspective from individual's mental factors, knowledge and instincts (Hodgson, 2006, p3-4; Polanyi, 1967).

Nevertheless, it is worth noting that, some of these views have certain reference significance for today's institution studies (Gross, 1999). Firstly, the scholar of the Behaviorism School argued against the proposition what the New Classical School insisted that economic reality was always static and unchanging—capitalism was a natural, reasonable and perfect system. Instead, Veblen (1899, p188) he believed that the emergence and innovation of institutions were an evolutionary process, rather than in a static and stable status (as the old institutionalism as claimed) (see Gross, 1999, p33), and humans were able to change social order in accordance with their own preferences[1]. Secondly, Commons realized that order and rules were generated by the individual interactions that originated from the process of collective selections (1934, p648). He

---

[1]   Veblen, 1919, p8-9, The Vested Interests and the Common Man ,"The habitual disuse which so allows the ancient canons of knowledge and belief to fall away…, is reinforced by the advancing discipline of a new order of experience…, and thereby brings on a revaluation and revision of the traditional rules governing human relations. The new terms of workday knowledge and belief, which do not conform to the ancient canons, go to enforce and stabilize new canons and standards, of a character alien to the traditional point of view. It is, in other words, a case of obsolescence by displacement as well as by habitual disuse"

described them as the collective actions in control of individual actions (ibid.). It means, the institution itself was created for some kind of collective purposes, rather than that were referring to by Adam Smith (1759, p380-381)[2], Carl Menger (1883, p130, p133 and p146)[3] as a spontaneous order evolved from the naturally formed (or unplanned) ideas, or as the habits of social interactions derived from human instincts (e.g. thinking, feeling, acting human being).

Based on above two revelations—i.e. institutions are evolving and are the outcomes of collective choices, new institutionalism emerged since 1980s. The new institutionalists advocates "rediscovering institutions" (March and Olsen, 1984), and "bringing the state back in" (Immergut, 1998, p17; Evans, Reuschmeyer & Skocpol, 1985; Cammack, 1989). Simultaneously, it re-emphasizes on what was left out of the earlier institutional studies that some discussions on the informal institutions (Gross, 1999, p33). Represented by Hayek (1960), March and Olsen (1984), North (1981, 1990), new institutionalist has provided an expanded definition and understanding of the function and change of institutions in the political world (Hodgson, 2006; Compell, 2007).

In this book, I am more agreed with the definition of institutions that North (1981, 1990) proposed. He explained the institutions as

---

2   Smith,1759, p380-381, "Thus self-preservation and the propagation of the species are the great ends which nature seems to have proposed in the formation of all animals"... "Nature has directed us to the greater part of these by original and immediate instincts.")

3   Menger, 1883, p133, "Social institutions simply cannot be viewed and interpreted as the product of purely mechanical force effects; they are, rather the result of human efforts, the efforts of thinking, feeling, acting human beings"

"a set of rules, compliance procedures and ethical behavior norms" (1981, p201-202), which are "consisted of both informal constraints (e.g. conventions, sanctions, taboos, customs, traditions, and codes of conduct), and formal rules (e.g. norms, laws and property rights)" (ibid). As a whole organizational apparatus constituted by strict rules, cultural and social orders, institutions act as the guidelines of individual and organizational actions, and play a regulatory role in their interactions.

In this regards, this apparatus relates to: 1) *fuzzy boundaries*, defining who is able to participate in the particular political arena, and outlining the framework of activities and the behavioral arena (Scharf, 1989, p152-154, 1990, p484; Steunenberg and Vught, 2012, p17; Walliser, 2006, p3); 2) *decision-rules*, shaping the political strategies of various actors, and governing and coordinating interactions in and between these actors and organizations (Hill, 2001, p39; Scharf, 1989, p156, 1990, p484); and 3) *feedback mechanism*, influencing what these actors believe to be both possible and desirable (Steinmo, 2011, p560), such as the idiosyncratic preferences and expectations of individuals (Rhodes, et al., 2008, p3; Hall, 1994, p48), and how the actors will be punished in the case of breach of the rules (Belle, 2015, p25).

The emergence and change of institution—interpretations from institutional equilibrium theory and evolutionary game theory

As mentioned above, the established rules, customs as well as practices are rooted in a particular historical and cultural context, and meanwhile the contextual factors are always unstable and changing. Thus, institutions are best considered to be "a set of process rather than a static thing" (North, 1990, 1991). This inevitably triggers a series of questions, namely how institutions emerge, how they function, what are the incentives for their change, and so on. In order to answer these, this thesis draws on several concepts from both institutional equilibrium theory and evolutionary game theory.

Institutions are defined as a set of rules of the game in a society (North, 1990, p3. It is directly designed by the participants as formal guidelines of conducts through a collective choice process (Ostrom, 2005; Greif and Kinston, 2011, p15). Besides, institutions are also regarded as the outcome of the game itself (Schotter, 1981; Greif 1993). Here, the outcome of the game refers to a "self-sustaining system of shared beliefs" regarding how the game is played (Aoki, 2001, p10). At last, they concern about the incentive mechanism of players to follow the particular rules consciously (Greif, 2006, p11; Gleif and Kingston, 2011, p14). The "self-sustaining system of shared beliefs", also known as the "self-enforcing rules" (Gleif and Kingston, 2011, p14) or "self-enforcing patterns of behavior" (Gleif, 2006, p11), are formed in a silent way, through which the game is repeated played (Aoki, 2001, p10) and the beliefs have been continuously enhanced. The shared beliefs can be identified with equilibrium attributes (or rather a "coordinating device") of the game (North, 1991; Greif and Kingston, 2011). Once all new players voluntarily comply with the rules, or are forced to obey the existing rules as well as an incentive mechanism of behavior in the game (Schotter, 1981, p46; Greif and Kingston, 2011), it could say that "a temporary institutional arrangement has be established, or that the game reached a temporary equilibrium" (North, 1991). As such, institution can be understood both as the artificially designed rules of the game and

31

the self-enforcing rules. These two interpretations are not conflicting; instead, they have much in common and complement each other rather than substitute. The former "has been fruitfully applied to reveal the emergence and functioning of a variety of institutions" based upon a perspective of rational choice (Greif and Kingston, 2011, p15). And the latter enables a more satisfactory treatment of the understanding of changing processes (Greif, 2006).

Equilibrium is a short-term, but static state of institutions. An equilibrium solution of the game emerges from individual attempts at self-maximization (Schotter, 1981, p46). Institutional change is considered to be the self-interested behavior of individuals and organizations (North, 1990; Bell, 2002; North and Thomas, 1973; Williamson, 1975) who attempt to change the rules for their own benefit (Greif and Kingston, 2011). In this process, an exogenous shock is the key force that breaks the equilibrium of the game, and generates a period of uncertainty (Greif and Laitin, 2004, p634, p639). Then, the contests among rational, self-interested individuals and organizations, determine the next equilibrium. As North (1991) argued, the external environment changes will cause the endogenous and exogenous impacts on the existing institutions, which would thereby bring about an unequal distribution of profits between actors. Once one or several groups—each of them is constituted by a number of individuals who are bounded together by a common goal, culture as well as other social attributes (Olson, 1965; 1982)—are expected that their participations in breaking the existing institutional arrangements will make themselves profitable, they would undoubtedly be committed to promoting the institutional change. Through a "centralized process of bargaining and political conflict between the groups" (Greif and Kingston, 2011, p14), the old institutional framework might be broken, and new compromises in the form of new formal rules and beliefs are reached. The capabilities of controlling resources among different groups determine a final equilibrium state.

Drawing on the concept of evolutionary game theory, participants in a game are not always constant while they are coming and going at any time; furthermore, the game is not a one-off game (North, 1990, 1991; Schotter, 1981). With the newly involved players and repeated games, therefore, there may be many different alternative institutions and equilibriums (Aoki, 2001). Hence, institutions evolve constantly with the continuously recurrent plays of games, and the equilibrium jumps from one to another. Schotter (1981, p29) terms this repeated process as a "super-game".

In accordance with North's path-dependence theory, for any politico-institutional arrangements, there exist "self-inertias" in their interiors (North, 1990, p95). After equilibrium is broken, old institutional arrangements will not fully leave the stage; instead, they change their role to become one of the players attending in the game. Due to this, new institutions are more or less affected by the existing framework, and in the same way, they will influence the institutions that may occur in the future. Besides, each institutional system is followed by a sort of interest distribution structure. Once a set of institutional arrangements has been formed, it will spawn a group of vested interest groups. They have strong demands for consolidating the existing institutional arrangements, so as to protect their advantages of accessing to the resources and benefits. Therefore, their conservatism is bound to interfere with the direction of institutional evolution.

2.1.2 Urban governance in a particular institutional context

Governance can be understood as the science of decision-making. It refers to a complex set of processes that society uses to regulate, coordinate and control its development and to resolve conflicts (Pierre, 1999, p376) by virtue of the formal and informal contextual attributes, such as values, norms, rules and laws, culture, customs, and so on, i.e. institutions (or "governance culture") that are embedded in the society (Healey, 1997, p65; Hohn and Neuer, 2006, p293).

At first glance, "governance" was taken as a synonym for governing by governments. Here the term is explained, in a narrow sense, as the active actions of political authorities—that is, "top-down governance" (see Driessen et al., 2012), by which the authorities attempt to shape socio-economic structures and processes in compliance with their blueprints. However, this definition seems to be extremely limited in its explanation of the newly emerged governance models characterized by their less bureaucratic and web-hierarchical structures (Jessop, 2000, p16), in which municipalities and states are no longer as dominant as they used to be in actions.

More recently, the term "governance" has become a virtual synonym for public management and public administration (Frederickson and Smith, 2003, p225), which indicates the multiple relationships and roles among the state, market and society (Hohn and Neuer, 2006; Healey, 2006). According to Hohn and Neuer (2006, p293), urban governance refers to a set of "collective and institutional regulations of urban development process" anchored by a particular group of actors, such as decision-makers on various spatial dimensions. As these actors act in a dominant role, their values, norms, cultural cognitions, and politico-economic positions have a direct or indirect influence on the formation of the regulation framework (Hohn and Neuer, 2006). Through the regulation framework, all relevant actors get involved in a game and interact with each other in the form of competition, negotiation, and cooperation which have been established by the rules of the framework. Further, the form and intensity of their interactions are subject to the formal and informal, flexible and long-standing framework of regulation, as well as the power balances derived therefrom.

Core components of governance and the mechanism of interactions
Actors, organizational sets and institutions constitute the core components of the concept of governance (Belle, 2015, p21). The study of governance is primarily related to the relation network of actors derived from their temporary or persistent interactions, forms of regulation, and cooperation; the governance process of specific organization sets; and the broader institutional context in which the above components are embedded. Here, I would like to borrow North's classic example—a "competitive team sport" (North, 1990, p4-5), to briefly illustrate how actors, organizations, and institutions work together in the process. Borrowing his view, actors maneuvering in governance arenas can be considered as sports players in a team competition (Belle, 2015, p25). Institutions are equivalent to the rules of the game, consisting of formal written rules as well as unwritten codes of conduct (Ramstrom, et al., 2006), the latter of which could be seen as the supplements to the formal rules. They "define the way of the game to be played, including both what players are prohibited from doing and, sometimes, under what conditions some players are permitted to undertake

certain activities" (North, 1990, p4). Players act as participants in the governance process, and form a team (i.e. actors and organizations of governance) for a common goal, namely beating the opponent and winning the game (i.e. goal of governance). Within the constraints of the rules, the teams are free to develop their own strategies which will be thereafter realized through players' cooperation and adjustments (i.e. strategy and operation of governance) (Belle, 2015, p25).

In the real world, actors involve both individuals and agencies. Organization sets refer to: 1) political bodies, economic bodies, social communities, educational entities, and so on. (North, 1990, p5)—that is, a group of collectives combined purposefully by the actors either with similar social attributes/identities, sharing the same cultural values, or holding common goals/interests (see also North, 1990), and 2) the public-private, hierarchical and network relations derived from their interactions (Hill, 2004, p7). Either actors or organizations may act as the initiators of public actions and simultaneously participate in the process of governance. The roles they assume are largely dependent on their abilities to access resources and their positions in the power structures (involving the authority and legitimacy of publishing rules and agreements, and the capacities of coordination and operations). The set of "apparatus", constituted by the organization sets and multi-level relations, guides the actions of participants and establishes links between actors/organizations with different identities and interests. It thereby generates a possibility of "collective actions." Of course, the individual differences and diverging interests of actors will inevitably lead to conflicts and confrontations. Thus, the apparatus also aims to provide mobilization based on an interest distribution mechanism.

However, it is worth noting that, due to the unequal status in power structure, the influence of each actor/organization participating in the collective actions is different. That is to say, this collective action" is not established on the basis of that all participants' needs and prospects are well-balanced and treated equally. Some of the actors gain an upper hand by virtue of their advantage positions in the power structure. It makes them closer to the center of policy-making on responsibility allocation and resource and interest distribution. Thus, they have more initiative in setting goals, formulating rules, and taking part in the practices during the governance process, and they even make decisions inclined to their own preferences and interests. Others obviously fall to a subordinate position—that is, playing a relatively weak role and making limited choices within the established rules.

As for the widespread unbalanced power status in the governance process, Foucault (1988b) gives an impersonal explanation. He considers the power structure and the strategies that consolidate or break the existing balance as the outcomes of the dynamics of local cultural and institutional settings. Power in the governance process involves all forces' relations and exists between actors in the form of applying and being applied, but it does not belong to any specific entities. In this regard governance can be also understood as a process in which some entities control, regulate, and dominate others (i.e. applying power) while meanwhile serving as a feedback mechanism when the applied power encounters resistance and then makes timely strategic adjustments. The apparatus, following Foucault's (1988a) opinions, refers to the knowledge, strategies, (visible and invisible) technical instruments, and so on. This is exploited by the actors standing in the different positions in the power geometry. Nevertheless, owing to the widespread and unequal power relations, the design of the apparatus and the practices of governance tend to be affected by the preferences of a certain group of actors who are given advantages by the established rules of the game (i.e. institution).

This preference comes directly from the actions and prejudices of the actors (individuals/collectives), but in fact, at a deeper level, is derived from the specific cultural and institutional contexts in which they are embodied. Actors are the basic units who act as decision makers, implementers, and participants in governance. Their ways of thinking and the forms of their behaviors are largely subject to their cultural attributes and socio-political contexts. Accordingly, their choice of technical instruments and the way they use techniques as well as the effects of their implementation are dependent on the cultural and institutional backgrounds as well. According to Ross (Ross, 1997, p42), norms, rules, and values in social life act as the basis of social and political identity that affects how actors line up and how they act on a wide range of matters. Pierre (1999, p373) emphasizes the necessity of considering institutional aspects in the study of governance and compares it to "a vehicle for understanding the values and objectives" that give "directions, objectives, and meanings to the processes of public-private resource mobilization, and the arising interaction networks based on the resource flows." Hence, an analysis of the governance process is inherently concerned with the role of norms, rules, and values in social actions (Healey, 2006, p301) and with the "technical instrument" invented by actors embedded in a particular context—or rather the "conduct of conduct" (Foucault, 1988a) through which power acts on the action of others and governs individual's behaviors (Foucault, 1988a; 1988b). Based on this, two questions need to be answered, namely, 1) how do norms, rules, and beliefs in a specific social context facilitate shaping a distinct form of governance through sorts of social actions; and 2) what is the micro-mechanism through which the norms, rules, and beliefs is applied by a certain group of actors to a wide range of collectively social actions?

Both of the questions are related to the role of institutional arrangements in shaping governance. Here I am more inclined to agree with Schimank's opinion (2004, p293) that institutional arrangements play a role in governance by affecting "inter-subjectively shared normative orientations, cognitive patterns of interpretation, and evaluative orientations of action" (Hohn and Neuer, 2006, p293). The former (i.e. inter-subjectively shared normative orientations) refers to a set of legitimate rules which are purposefully designed and maintained by the dominant actors in the power relations to regulate actor's behaviors, constrain collective ethics, and affect the formulation of goals, interests, and strategies of all actors (Pierre,1999; Healey, 2006; Hohn and Neuer, 2006). By specifying who has permission to access to the collective actions and simultaneously to normalize and standardize the operating procedures at a legal level, formal institutions define the respective roles of participants in governance (e.g. the position in power relations); structure the relationships among different actors; orient the hierarchical networks and organizational structures that shape and support such interactions (e.g. administrative levels, social networks, and coalitions); and influence the policy-making of actors (e.g. framing issues, selecting solutions and formulating plans). All the above constitute a distinctive model of governance.

The latter two (i.e. the cognitive patterns of interpretation and the evaluative orientations of action), referring to beliefs, customs, and codes of conduct, affects the manner of governance through the collective cognitive, and spiritual mobilization, and self-discipline of actors. It is linked governance by ideological constructions (Ramsey, 1996, p79) through which participants in the collective actions understand social and political realities, interpret local events, and define socio-economic issues. Moreover, it makes policy decisions based upon the core values provided by the ideological orientations of the specific cultural and social setting (Difaetano and Strom, 2003, p360), and then guides the distinct socio-economic actions of actors in the process of governance.

Governance model interacting with institutional change
The formal institutional rules and cultural context are non-contradictory in evoking a new governance model. Instead, they shape the meaning that actors assign to objects, practices and places as well as to their own identities (Etzold et al., 2012, p187) on the one side, and they normalize actors' behavior and practice procedures on the other side. Through constantly reshaping the consciousness and behavior of actors, institutions provide persistent forces in promoting the formation and transformation of governance, in which social practices act as an intermediary (Figure 2.1). In other words, certain institutional and cultural features are reflected through actors' social practices with subjective preferences, and meanwhile, their behaviors, interactions and policy orientations in the practices constitutes different patterns of governance.

Healey (2004, p14) calls the governance setting, which is largely affected by the institutional context that social practices are embedded in, the "mobilization of bias". It is also termed the "governance culture" and "planning culture" (Hohn and Neuer, 2006, p293; Healey, 2006, p310). This "mobilization of bias" is variable and changes along with institutional transition but is stable over a period of time. Based on the "bias", strategic projects for governance purposes are created (i.e. the decision making process), coalitions and the balance of power are built on the bais of the actors' role, responsibility and power positions (i.e., organization setting and networks) are established, and the way of power operation are played out (i.e., the implementation process). The decision-making and implementation process embedded in a particular governance culture, or further rooted in a wider scale institutional context in society lead to the flows of public and private resources—i.e. the distribution and exchange of material resources, capital, rights, obligations, and interests—as well as the flexible and temporarily stable communication networks to support the flows (Figure 2.1).

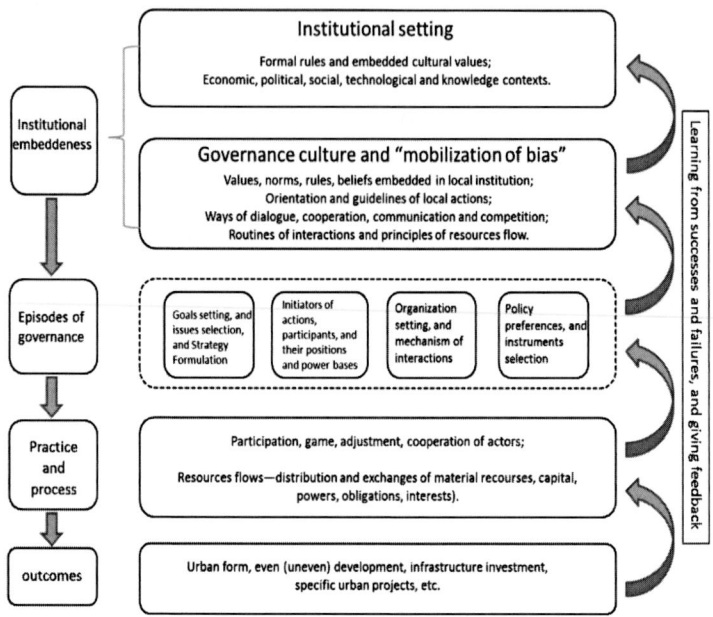

Source: Hohn and Neurer, 2006, p 294; Healey, 2006, p305-306, with modification

Figure 2. 1 Urban governance with institutional embeddedness

The governance cultures and governance process (i.e. decision making, pattern of communication and implementation procedures) are not always stable and unchangeable. In fact, once an institutional system in a particular socio-economic context is changed (along with the game among different interest groups in response to the economic, political, social, technological and value-related transformations), a set of power relations inherent in the system will simultaneously be restructured. Accompanying the restructured power structure, the dominant actors are replaced and a set of new policy-making rules related to resource distribution and responsibility allocation, ways of cooperation and participation, instruments selections, and so on will also change successively. It thus gives rise to a new pattern of governance. In turn, the governance process will provide potential driving forces to a wider scale institutional transformation in a society through summaries of or reflections on the successful/failure experiences of actors according to the strategies, power- and interest-based relations, policy configuration, and so on in their collective actions. Furthermore, one thing should not be ignored, as Foucault argued (2008) that governance is supposed to be viewed as a process of dominating power playing a role in disciplining individuals. Where there is power, there must be resistance in the process. Following this idea, governance should be also considered as a process in which different actors contest with each other. Along with their competition and compromise, the goals, policy configuration, organization setting, and thus derived interactions have the potential to change. It therefore challenges the original governance model and governance culture and exerts further influence on the broader institutional background.

Model of governance

As described above, different governance models are originated in the corresponding institutional contexts, and are distinguished from each other by virtue of distinct participants, primary objectives and preferred strategies, institutional foundations, main instruments, the most common outcomes and so on (Driessen et al., 2012, p145-148; Pierre, 1999, p377). According to the differences of above components, Driessen et al., (2012, p145-148) have summed up four general models of governance, namely central and decentralized governance (or managerial governance), public-private governance (pro-growth governance), interactive governance (corporatist governance), and self-governance (Figure 2.2).

In *central and decentralized governance,* either central or local government takes the lead role, and the market and civil society are the recipients of that government's incentives (Lin, Hao and Geertman, 2015, p1777). It emphasizes professional participations over elite political involvement (Pierre, 1999, p380), whose main purpose is to enhance the efficiency of public service production and delivery.

Public-private governance is characterized by the joint actions of partners in public and private sectors, by structuring the concerted public-private actions to boost local economy (Lin, Hao and Geertman, 2015, p1777; Pierre, 1999, p384). The shared interests between municipal authorities and business elites act as the most fundamental forces to promote such collaborative and mutually beneficial actions. National traditions of state strength and public presence in the markets provide the political and social preconditions for such cooperation (Savitch, 1998; see Pierre, 1999, p384). In the public-private governance, private actors are more probably granted high priority in collaborations. It, therefore, leads to "a privileged position of private capitals" in urban policy formulation, and the process of development (Pierre, 1999, p384).

Interactive governance emphasizes solving societal problems and creating societal opportunities through interactions among the state, private sectors and civil society (Kooiman et al, 2008, p2). In this mode, actors from civil society, public and private agencies are interacted with each other in the form of networks. Each of them is

44

in the equal status in collaborations (Driessen et al, 2012, p146; Lin, Hao, Geertman, 2015, p1777).

Self-governance is characterized by far-reaching autonomy that is enjoyed by stakeholders from the market and civil society (Lin, Hao, Geertman, 2015, p1777). The involvement of stakeholders and their roles are determined by market and society, and their interactions are closely linked with deliberations and negotiations.

		Centralized governance	Decentralized governance	Public–private governance	Interactive governance	Self-governance
Actor features	Initiating actors	Central gov't agencies (or supranational bodies)	Gov't at its various levels of aggregation (subsidiarity)	Central gov't agencies; private sector is granted a preconditioned role also	Multiple actors: gov't, private sector and civil society	Private sector and/or civil society
	Stakeholder position	Stakeholder autonomy determined by principal agency	High likelihood of stakeholder involvement	Autonomy of market stakeholders within predetermined boundaries	Equal roles for all network partners	Self governing entities determine the involvement of other stakeholders
	Policy level	(Supra)national state	Lower levels of gov't	Local to international level	Multiple levels	Local to international level
	Power base	Coercion; authority; legitimacy (democratic representation at the national level)	Coercion; authority; legitimacy (democratic representation at lower levels)	Competitiveness (prices); contracts and legal recourse; legitimacy (agreement on relations and procedures)	Legitimacy (agreement on roles, positions, procedures and process); trust; knowledge	Autonomy; leadership; group size; social capital; legitimacy (agreement on relations and procedures)
Institutional features	Model of representation	Pluralist (popular (supra)national election and lobbying)	Pluralist (popular local election and lobbying)	Corporatist (formalized public–private governing arrangements)	Partnership (participatory public–private governing arrangements)	Partnership (participatory private–private governing arrangements)
	Rules of interaction	Formal rules (rule of law; fixed and clear procedures)	Formal rules (rule of law; fixed and clear procedures)	Formal and informal exchange rules	Institutions in its broadest form (formal and informal rules)	Informal rules (norms; culture); self-crafted (non-imposed) formal rules
	Mechanisms of social interaction	Top down: command and control	Sub-national governments decide autonomously about	Private actors decide autonomously about collaborations	Interactive: social learning, deliberations and negotiations	Bottom up: social learning, deliberations and negotiations

Literature Review and Theoretical Frame

Features concerning content					
Goals and targets	Uniform goals and targets	Uniform and level specific goals and targets	Uniform targets goals; specific actor (collaborations within top-down determined boundaries / determined boundaries)	Tailor-made and integrated goals xand targets	Tailor-made goals and targets
Instruments	Legislation, permits, norms and standards	Public covenants and performance contracts	Incentive based instruments such as taxes and grants; performance contracts	Negotiated agreements; trading mechanisms; covenants; entitlements	Voluntary instruments; private contracts; entitlements; labelling and reporting
Policy integration	Sectorial (policy sectors and levels separated)	Sectorial (policy sectors separated)	Sectorial (branches and industries separated)	Integrated (policy sectors and policy levels integrated)	Sectorial to integrated (depends on problem framing by communities of interest)
Policy-science interface	Primacy of generic, expert knowledge	Primacy of generic expert knowledge; room for issue and time-and-place specific knowledge	Dominance of issue and time-and-place specific knowledge; expert and lay (producers and consumers)	Transdisciplinarity: expert and lay knowledge in networks; emphasis on integrated and time-and-place specific knowledge	Dominance of issue and time-and-place specific knowledge: expert and lay (citizens)

Source: Driessen et al, 2012, p146-147.

Figure 2. 2 Modes of Governance

2.2 Neoliberalization—a worldwide steering since 1970s

2.2.1 Raising to a prevailing ideology in the global

In the last four decades, there has everywhere been a new, emphatic institutional shift towards neoliberalism—i.e. a resurgence of ideas associated with laissez-faire economic liberalism—in both political-economic thinking and practices.

Keynesian failure in the global economy

As a newly institutional mechanism, the rise of neoliberalism is a response to the economic instability, high inflation and mass unemployment of the 1970s, and in face of ongoing serious economic crisis, Keynesianism became helpless (Brenner and Theodore, 2002, p350; Harvey, 2005, p1-2; p12). This theoretical framework in economics and economic policy-making that had been fashionable at a time between 1940s and 1970s was no longer to be trusted by many countries that national macro-controls on finance, tax and employment would keep on producing advantages to profit-making and economic growth.

In the neoliberal view, construction of the market system in accordance with Keynesianism has its inherent drawbacks. For instance, the economic policy expansionary in response to the inflation that Keynesian states pursued—i.e. stimulating economy and maintaining prosperity through the expansion of government spending, the implementation of the budget deficit and increased demand—cannot solve the stagflation, but seemingly makes the problem worse (Harvey, 2005, p8-10). Even so, this initiative had successfully helped the Western capitalist countries tide over the Great Depression of the 1930s.

The neoliberal scholars represented by Milton Friedman and Friedrich Hayek criticized harshly Keynesian interventions and/or collectivist strategies (Beatty, 2014, p42; Peck and Tickle, 2002, p381; Brenner and Theodore, 2002, p350). They simultaneously argue that the economic turmoil of capitalism happened in the late of 1970s should be completely attributed to the initiatives of government's intervention in economic fields, excessive public investments, over-concerning about employments as well as a series of wrong financial, monetary and tax policies. Meanwhile they believe, only liberalization and marketization can help the states successfully avoid the ills derived from excessive re-accumulation of capital reproduction (Jessop, 2002, p454-456; Dumenil and Levy, 2004, p2). To this end, they put forward three core principles, namely the fully private property, unrestricted trading and investment, and undisturbed market, which were bonded with "aggressive forms of state downsizing, austerity financing as well as public service reform" (Peck and Tickell, 2002, p381).

Financialization in the global since 1970s

If the economic stagflation and Keynesianism failure in economic recovery and social stability since the early 1970s provided an initial impetus to re-examine the government intervention and free market, then the growing role and power of finance in the political economy of capitalism (Kotz, 2008, p1), and the shift in the competitive structure of global capitalism (i.e. widespread international competitions), should be considered as the root causes of putting neoliberal ideology into actions. Jessop (2002, p455) deemed the resurgence of liberalism in the form of neoliberalism as a successful hegemonic project voicing the interest of financial and/or transnational capital. In this regard, Dumenil and Levy (2004, p1-2) gave a further explanation: "neoliberalism is the expression of the desire of a class of capitalist owners and the institutions in which their power is concentrated, which we collectively call "finance", to restore...the class's revenues and power..."

Entering the 1970s, financial markets started to be active, and the profits of financial institutions increased substantially compared to the growth rate of non-financial sectors (Kotz, 2008). In addition to the quantitative growth of financial activities, financial institutions start to turn their sights on the non-financial sectors. According to Kotz (2008, p5), these institutions, in particular referring to the financial giants, have abandoned their role as servants of non-financial sectors, and are more committed to the innovation and marketing of diverse financial instruments and products.

In essence, financialization is the reshape of relationships between different economic groups under capitalism. It is an outcome that financial capitals winning over other economic sectors (e.g. production capitals) in both domestic and international economic environment (Kotz, 2008, p7). In face of the strictly regulatory policies and measures in Keynesian period, the voices—i.e. getting rid of financial restrictions, creating relaxed monetary environment, eliminating investment barriers internally and internationally, and regaining a dominant position of financial capital, as Kotz (2008) argued—grew louder and louder among financial oligarchs.

In this context, neoliberalism came to the fore based on a series of bargaining and political competitions among the interest groups. Its appearance is the product of the state's efforts (represented by ruling elites) to further advantageous conditions for domestic and international capital investments (Harvey, 2005, p19; Kotz, 2009, p316), which is precisely what the capitalist class expects to see. Hence, it could be said, in some sense, that neoliberal policies stood out as a perfect "political and strategic partner" in response to constantly growing power of financial capital (see also Harvey, 2005; Kotz, 2008, 2009). It is an "outgrowth" of neo-capitalism characterized by the rise of financial capitals, which is aimed at removing all "roadblocks" for global economic integration.

Extensive attempts around the world
The neoliberal doctrine first took root in the USA and UK (Kotz, 2008), and was then quickly accepted by most of Keynesian welfare national countries as well as some developing countries, even by several communist states that were facing economic recessions in the global crisis at the time. At the end of 1970s, the United States and UK took the lead in succession in implementing the "anti-interventionist" policies to revitalize their depressed economies (Harvey, 2005, p31; Kotz, 2008, p6). These policies relate to containing the powers of labor, relaxing the development and extraction rights on industry, agriculture and other resources, cutting taxes for business and wealthy individuals, and liberating the capital and finance on both domestic and world's stages (Kotz, 2009, p307; 2008, p11).

Then, in face of serious domestic inflation, debt crisis, and the social conflicts resulting from rapid declines in individual earnings and income, a series of liberalization reforms measures stemmed from the projects of "Baker's Plan" and "Washington Consensus" were experimentally applied to Latin America in the late 1980s (Howard and King, 2008, p102). These measures involved in: 1) the trade liberalization and privatization of state enterprises; 2) reducing or even abolishing national interventions and controls over prices, exchange rates, rents, wages and so on; and 3) establishing financial markets and simultaneously relaxing restrictions to foreign investment.

The third landmark event of neoliberal shift during this period was the collapse of the communist regimes of Eastern Europe. A range of radical reform programs—the so-called "Shock Therapy" which was estimated to be quicker, exhaustive and efficient than gradual approaches, were widely adopted in these countries. After undergoing reform, state's intervention was completely negated, national public sectors and social resources were largely privatized, the domestic door of these countries was totally opened to ready for welcoming international capital investment. Besides, the neoliberal idea quickly spreads to South Africa in the late 1990s, and as Harvey (2004, 2005) and Wu (2010, p619) argued, even in contemporary China—the state has always been grasping the absolute discourse right in the past, as we shall see, its reform appears to be headed in this direction.

2.2.2 Neoliberal localization—from a utopia to a geographic explanation

As Peck and Tickell (2002, p380) argued "neoliberalism does indeed seem to be everywhere". It is spread both as Beck (2000, p122) described "an ideological thought virus", and as Bourdieu and Wacquant (2001, p2) argued "a new planetary vulgate". Nevertheless, there is a growing consensus of opinion on its origin. It is widely viewed as a successful "encompassing hegemonic project" involving politics, economics and culture emanated from the "ideological heartlands" of the United States and the United

Kingdom (Hoffman et al., 2006, p10-11; Peck and Tickell, 2002, p382; Larner, 2003, p510), and has rose up to "a common-sense of new world order" to shape our world today (Peck and Tickell, 2002, p381)

Broadly, neoliberalism serves as a philosophy ranging over a wide expanse in regard to ethical foundations as well as to normative conclusions (Blomgren, 1997, p224; see Thorsen, 2012, p185). It is best understood as "a loosely demarcated set of political beliefs" (Thorsen and Lie, 2007, p14), which roots in the faith of "individual freedom", and then gives a high priority to civil and economic liberties, and the right to private property at theoretical spheres. However, as Brenner and Theodore (2002, p16) point out, neoliberalism contains not only "a utopian vision of a fully commodified form of social life", it simultaneously represents a "complex, multifaceted action involving socio-spatial transformation". It refers to a concrete program of institutional modifications aiming at creating an "unfettered rule" for capital flows (Brenner and Theodore, 2002, p16; Dumenil and Levy, 2004, p2, see Kotz, 2008), and a new form of capitalism, which controls and dominates individual activities through rules-making, knowledge dissemination and practices (Foucault, 1977; Lemke, 2000; Ong, 2006).

The associated ideology has penetrated into many practical fields. However, it could be more obviously to be noticed in economic and policy-making spheres. Some neoliberals represented by Friedman, Nozick, Rothbard and Hayek praise highly the "anarcho-liberalism" (Thorsen and Lie, 2007), and argue for a "complete laissez-faire, deregulation and the dismantlement of all government" (Blomgren 1997, p224, see Thorsen and Lie, 2007).

This view of the so-called "roll the state back" was dominant during the early phase of neoliberalism. After that, some other scholars advocate a principle of "minimum state" in contrast to the "big government" of the Keynesian period (Harvey, 2005, p2; He and Wu, 2009). Under this argument, states/governments with functions should exceed those of the so-called "night watchman" states, whose primary responsibility is playing limited role under the premise of a full market (Thorsen and Lie, 2007, p12).

After entering the 1990s—i.e. the "roll-out" era defined by Peck and Tickell (2002, p384), the intervention of state/government has not been entirely negative as before, but are re-examined by neoliberals. The "active state-building" with selective interventions, involving the purposeful consolidation of "new state forms, modes of governance, and regulatory relations are emphasized" (Jessop, 2002, p454; Peck and Tickell 2002, p384). The purpose is to guarantee a smooth running market, and thereby improving the competitiveness and economic efficiency. However, the regression of intervention is fragile and temporary (Jessop, 2002, p454). This form of regulatory is more akin to a "hyper-marketed style of governance" (Weber, 2002, p520). That is to say, the core objective of all government behavior is to serve market, and meanwhile to exercise duties in the framework of market rules. Once the transition ends, the state will retreat to its minimal role (Jessop, 2002, p454).

Although debates around the relationship between state and market are ongoing, the perspectives from different disciplines have reached a consensus. That is, this doctrine advocates highly free and competitive market; "free order" in market; the limited involvement of government; and a set of public-private partnerships instead of "top-down" arrangements in Keynesian era. Specifically, it relates to:

1. Enhancing the role of the market. Creating any possible conditions to the free movement of goods, services, and especially capital, throughout the global economy (Kotz, 2008, p3). For instance, to liberate private enterprise from any imposed government association, and opening doors for cross-national/ cross-regional trade and investment, as what authoritative international agencies (e.g. WTO) prescribes.

2. Limiting government role. Reducing government intervention in economic activities as much as possible, especially in the field of capital investment, as well as financial and monetary organizations (Martinez and Garcia, 1996). Instead, encouraging any forms of self-organization, and the networks and partnership relying on consultation and negotiation are promoted (Jessop, 2002, p460; Brenner and Theodore, 2002, p365).

3. Praising the dominant role of capital. Cutting public expenditure for social services, and canceling state social program (Brenner and Theodore, 2002, p364); making social policy be subordinated to economic policy (Jessop, 2002, p459); shifting a cooperative relation between capital and labor into a kind of unbalanced affiliation, in which labor are dominated by capital with the aid of state.

4. Enhancing inter-capital competition. Withdrawing state from its role of supporting leading national industries (Brenner and

Theodore, 2002, p364), and replacing co-respective behavior by unrestrained competition of large corporations (Kotz, 2008, p3).

In the fields of economics and political science, neoliberalism has become a rather general concept, and the controversy around the proposition of "how neoliberal ideology is conceived and imposed" are rather fierce and ongoing. Similarly, in the spheres of geography and sociology, with reference to the widely divergent policies and experiences of the past 30 year in many countries in the world, the issue about "why different patterns of neoliberal practices are evolved, or are mutated at a range of geographical scales" (Ong, 2006), has also aroused a wide attention of related scholars.

Currently, it is easy to discovery the footprint of neoliberalism in every corner of the globe, and in various socio-political contexts. Regardless of the traditional democratic states (e.g. the UK and the United State), the social welfare states (e.g. Sweden, Norway and New Zealand), or the Post-soviet countries have been, to a certain extent, affected by the neoliberal thoughts since the end of last century (Harvey, 2005; Brenner and Theodore, 2002, p350). Besides, it is not "necessarily incompatible with state authoritarianism when it is viewed as a political and economic project" (Wu, 2010, p622). Some cases have been demonstrated in many literatures that many of these countries are more or less along the trajectory (Lin and Zhang, 2015, p2775; Wu, 2010, p58). As Harvey (2006, p34) observed: "... much of East and South-east Asia—in South Korea, Taiwan, and Singapore most noticeably—this connection between dictatorial rule and neoliberal economics had already been well established".

The policies and strategies derived from same ideology but adopted in the context of distinct regimes are quite different. Brenner and Theodore (2002, p368) term it as "the process of neoliberal localization". It refers to the distinct sets of policy combinations as well as concrete practices, with particular local-based forms and effects, within diverse socio-political contexts. The differentiated contexts exert a subtle influence on the scope and orientation of neoliberal shifts (Brenner and Theodore, 2002b, 2002a; Jessop, 2002; Peck and Tickell, 2002). Each variation of neoliberalism is largely dependent on its "historic traditions, development paths, and changing economic and political conjectures" (Eick, 2006, p68).

Following Swyngedouw's (2004) definition of "glolication", each variation of neoliberalism, might be more accurately understood as an interlinked outcome of the global ideology and local reality. In this process, the evolution of neoliberalism moves simultaneously towards two directions—that is the "upward" spread from the "heartlands" to peripheries (see also Peck and tickell, 2002), and the "downward" penetration and application (with various possibilities of mutation) at local (Swyngedouw,1997; Eick, 2006, p68; Brenner and Theodore, 2002). In the process of the "forward and reverse" flows, local practices and the values of this ideology are interacted with each other. This interaction therefore promotes an ongoing evolution of neoliberalism.

In addition, many geographical scholars assert that the "localization of neoliberalism" does not happen overnight. On contrary, it is more likely a fumble process—characterized by "open-ended" and "trial-and-error" (Brenner and Theodore, 2002, p360). Neoliberal turning is an outcome of the repeated struggles, consultations, adjustments and adaptations of stakeholders, and capital force as one of the stakeholders seems to play a crucial role. It is simultaneously an ongoing process rather than staying in a stable state once a compromise being reached. Thus, neoliberal shift in the real world does not always occur by a peaceful way. At national and regional scale, as Harvey (2005, p3) argued, it sometimes happens voluntarily, and in other instances are in response to coercive pressures. For instance, the shifts "in the US and the UK were progressed in a democratic manner"; while in Chile and Argentina, the reform was imposed by traditional upper classes as well as military force (Harvey, 2005, p39). While on a larger geographical scale, the "recognition on neoliberal beliefs and the observance of rigid rules" are imposed by some powerful financial institutions as Hackworth (2007, p18) referred to the "policy officers" in global, such as the International Monetary Fund (IMF), the World Bank and the Trans-Pacific Partnership (TPP). These are regarded as a universal value and a core principle that their members are forced to comply with.

2.3 Neoliberalism, capital accumulation and space

2.3.1 Nature of neoliberalism—facilitating capital accumulation

As mentioned in Chapter 2.2, the common nature of a variety of neoliberal strategies adopted by states with distinct institutional configurations is to facilitate an unfettered rule of capital circulation in wide spheres of production, and between different countries/regions in response to the crises and economic instability that have arisen since the mid-1970s. Many neo-Marxists have come to embrace that the neoliberal turn signifies "the monopoly stage in capitalism" (Kotz, 2008, p3) or, the new "social order"

(Dumenil and Levy, 2011; Parker, 2012, p196), or the "neoliberal phase of capitalism" (Sotiropoulos, 2011). Harvey (2005, p2, p19; Kotz, 2008, p12,) emphasizes that the neoliberal doctrine has seemingly become a successful "political-economic project (involving both practices and thinking) to diffuse the crisis and recession, to re-establish the conditions for accumulation, and to restore the power of economic elites." Its purposeful design is to provide a platform for stabilized and reproducible capitalist growth in the contemporary era.

The predominance of the ideology argued by neoliberals and capitalists is that it proposes a solution with immediate results to cure flagging capital accumulation, about which other theoretical hypotheses and ideologies (e.g. Keynesianism, neo-corporatism, neo-statism, and neo-communitarianism) are assessed to be ineffective and powerless (Jessop 2002, p460-461). The world mainstream doctrine presents clear criterion (i.e. deregulation, privatization, marketization, and so on.) on how and to what extent to rescale the capital-state relations, capital-labor relations, and capital-capital relations (affecting the form of competitions on both national and international scales) (Kotz, 2007, p7; Weller and O'Neill, 2014, p112).

In line with the universal criterion, both traditional neoliberal states (e.g. the USA and the UK) and the countries (e.g. Canada, Japan, Russia, Chile, etc.) that purposefully adopt or are imperceptibly affected by the ideology have devoted themselves more or less to an "institutional fix" in the international or domestic realms in the recent decades. These countries dismantle the inflexible and rigid "regulationist" modes—as Jessop (2005, p19-42) termed it—in their isolated territories. Simultaneously, they are working on creating a more open and market-based form on a much broader geographic scale (Kotz, 2008; Harvey, 2005). The obvious purpose of doing so is to clear all obstacles to capital circulation and to create a wide cooperative and competitive environment to accelerate this circulation.

The institutional fix serves to facilitate capital accumulation, and on this point, it can be noticed in the following specific aspects. At first, liberalization and narrowing state power to a minimum range (which is quite common, at least in the economic sphere) is proposed in the hope of securing a smoothly operating capital market and flexible resource and labor markets. Liberalization helps to remove potential obstacles for the free flows of capital in different production spheres as well as economic sectors, and on various geographical scales.

Second, another important doctrine of neoliberalism is that privatization favors opening up new fields for capital accumulation. By clarifying the property rights of public resources (e.g. mineral resources, water, electricity and transportation), social welfares (e.g. employment insurance, health insurance and pension), and public institutions and organizations (e.g. universities, hospitals, research centers and municipal libraries), capital has the chance to access the fields legally (Harvey, 2005; Hasenfeld and Garrow, 2012). Regarding to these, however, privatization was once shut out because of national/local protection policies.

Third, the principle of market-oriented allocation of social resources and wealth and of weakening governments' power in markets, as neoliberals support, is seen as a means for avoiding opportunities for favoritism, and also for reducing inefficiency and the potential costs associated with administrative interventions. On the one hand, the free market paves the way for the convenient exchange of resources, labor and money (Harvey, 2004). It favors flexible capital investments in the form of frequent and low-cost purchases and sales, as well as considerable profit-surplus generated therefrom. On the other hand, the powerful voice in creating a sound market and preventing any form of intervention, particularly the deregulation of the interest-rate market, has helped such a "financial system become one of the main centers of redistributive activity" relying on its solid capital strength (Harvey, 2004, p34). Afterwards, by inventing flexible and diverse approaches, the system is further committed to capital accumulation. For instance, by investing in the credit products and stocks that are offered by financial institutions, capital as well as wealth is arbitrarily brought together from individual investors (Harvey, 2004, p34); similarly, through which the generated earnings are allocated and returned to the investors according to the interest rate.

2.3.2 Space with new meanings in neoliberal era—from a perspective of "accumulation by dispossession"

Since the essence of neoliberalism is to serve capital accumulation, then in studying urbanization, urban development and city transformation in the neoliberal era, it is impossible to bypass two important issues, namely the nature of capital accumulation, and the importance of space in the accumulation process.

The nature of capital

The most "primitive and original" nature of the rise of capitalism, as Marx argued in his book *Capital* is in search of the continuation and proliferation of accumulation practices (Harvey, 2004, p32). According to Marx's opinion, capital accumulation relies heavily on the production of surplus value, as well as the ability of capitalist class appropriating surpluses, and launching the surpluses into circulation in pursuit of further surpluses. Once capitalist system is formed, struggles over the occupation, control, reinvestment of surpluses of the bourgeoisie in the circulation will not cease.

In his view, capital by its very nature is to ceaselessly seek to transcend all "boundaries" of market, on which surplus values associated with new markets are produced and appropriated (Marx noted in his *"Economic Manuscript in 1857-1858"*). The profit-driven nature will inevitably lead capitalist to try their best to extend powers to territories, sectors and domains in which surpluses are not yet incorporated into circulation of capital (Harvey, 2004, p71), or to those with relatively favorable natural conditions for the production of surpluses. If possible, it is pleasure for capital to make all visible and invisible resources (e.g. labor forces, land, cultural heritages and local customs) to be commoditized. Commercialization, as Marx defined, is the primary step for capital accumulation, through which surpluses are produced by labor force and appropriated by the capitalist class.

Space as a key production resource
But in *Capitalist*, space is excluded by Marx from the category of commodity. Space, the so-called nature defined by Marx is only treated as a place carrying/providing the raw materials of production, and as a meaningless "box" being ready for endless capital accumulation. In the "spatial box", capitals are invested in labors and raw materials, thereafter labors process materials and

produce surplus value, and then capitalists appropriate the surplus and invest them into production again. That said, all events regarding the production of surplus value and capital accumulation "occur on top of a pin rather than in space" (Richard Peet, 1991, p179). Many neo-Marxist scholars (e.g. Lefebvre, Harvey, Soja) and post-modern geographers (e.g. Elden and Foucault) argue that Marxism is not particularly noted for its attendance to questions of space for a variety of possible reasons (Elden, 2007, p 107). Historical accumulation, in its analysis of the capitalist mode of production, has been sufficiently valued, while geographic variation is always treated as an "unnecessary complication" and is wrapped in various simplistic assumptions (Harvey, 1985, p141; Harvey, 1982, xii; Soja, 1989, p32).

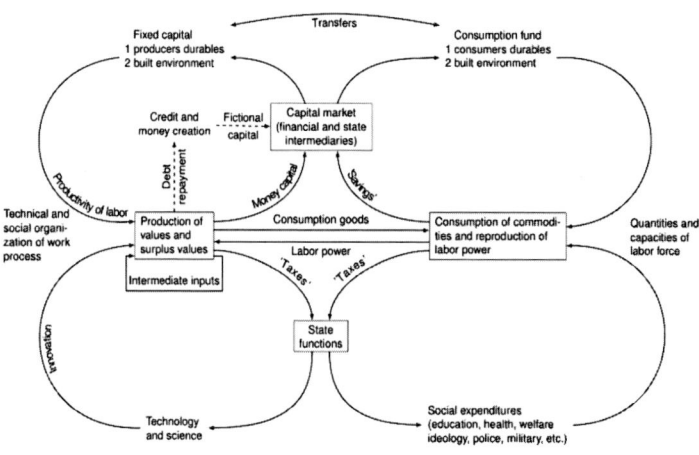

Source: from Harvey, 2006, p 208.

Figure 2. 3 Paths of capital circle

Despite Marxism has little to say about relations between accumulation and nature (Peet, 1991, p179), one of the revelations that Marx brought to us is that, his treatment of nature is not isolated from social production and capital accumulation; instead, nature is used in a variety of ways. In the derivation of the concept of commodity, Marx depicted capital accumulation as a process by which the form of nature is altered (Smith, 1984, p371). Marx's study "leads firmly in the direction that the development of the material landscape presents itself as a process of the production of nature" (Smith, 1984, p368). He recognized that capital accumulation took place in a historical and geographical context, which in turn led to specific kinds of physical landscape on earth (Harvey, 2001, p237). However, in Marxist writings, there is lack of comprehensive explanations on how such social reproduction under capitalism changes nature (space).

Inspired by Marx's definition of the capitalist mode of production in nature—that capital is always keen to expand the "boundaries" of market and to be in search of new material means of production on earth, neo-Marxist geographers (Lefebvre, 1976, 1984, 2003; Castells, 1972; Harvey, 1985, 2001, 2004; Smith, 1984, 2007; et al.) confer a whole new meaning towards nature/space under contemporary capitalism. By proposing the model of capital circuit (Figure 2.3), Harvey (1985, 2006) puts forwards that capital will inevitably switch its enthusiasm from investing in industrial sector, to urban environment as well as infrastructure, and then to social welfare (education, health care, etc.). "Capital switching" between different circuits of capital is viewed by Harvey as a fix to address the excessive accumulation.

In line with Harvey's concept, Lefebvre considers that "production no longer occurs merely in space under contemporary capitalism; instead, space is itself now being produced in and through the process of capitalist development" (Lefebvre, 2009, p185). In this regards, space is neither considered as a term of "nothingness", nor as a place/"natural box" that is only responsible for providing subjects, objects and instruments for social reproduction. On the contrary, it has been materialized after capital crisis (Lefebvre, 1973, p141-160; Soja, 1996, p74-75), and presented as a commodity that can be exchanged with capital (see Brenner and Elden, 2009, p130, p214). Harvey (2001, p235) in his book "*Spaces of Capital: Toward a Critical Geography*" supplements that, "it is an indisputable fact that capital and perhaps labor surplus are absorbed through they entering new territories in the form of geographic expansion and constructing a new set of spatial-economic relations". Once space is branded with the attribute of exchange, it naturally becomes a means of production, of course, a new investment venue for absorbing the excess capitals and labors, and for generating surplus value. As we have seen, from outer space to the sea, air, water and light in global, even the city itself as well as the spatial webs of communication and exchange have nowadays been favored by capital investment, and performed as important components to maintain the vitality of capitalist (Lefebvre, 1991; see also Brenner and Elden, 2009, p24).

The new interpretation of space in the contemporary capitalism, as Smith (1984, p91) argued, helps to "unravel the mystery of how capitalism attenuates its inner contradictions and defuse the ensuing economic crisis". It gives a possible answer on the issue of why capitalism has survived so far after Marx presented his prediction in the book *Capital* that "capitalism will eventually perish". In order to further reveal the process, Harvey and Smith propose a theory of uneven geographical development. Following with Marx's opinion that the essence of the "self-survival of capitalism" is to keep on maintaining a hierarchical structure through uneven development (Wallerstein, 1974; Dunn, 1998), which is characterized by the unequal market exchange, uneven appropriation of production materials, and uneven development of capital's productive forces in various regions, sectors and departments (Clarke, 1994, p77; Smith, 1984, p189; Radhika Desai, 2015, p137, Theoretical Engagements in Geopolitical Economy, Emerald Group Publishing), Harvey (2004, p55-89) and Smith's (1984; 2007) put forward a further insight: in the era of neo-capitalism, especially in the context of neoliberal strategies of globalization, uneven geographical development has turned into a necessary means to maintain the capitalist mode of production. Likewise, the complex physical landscapes at different geographical scales also confirm the results of capital accumulation in history.

First, producing space is a prerequisite for the basic law of capital accumulation what Marx's so-called "the annihilation of space through time" (Harvey, 2004, p76), and "the annihilation of time through space" proposed by Harvey (Harvey, 2004, p77). By introducing the concept of "spatial fix" (Harvey, 2001, p23; 2003), Harvey argues that entering the 20th century, the mode of capital accumulation has being involved in urban development and real-estate sector. Currently, increasing capital are invested in the built environment for a prolonged period, by which the material body of fixed capital is fossilized in the land scape to deal with the excessive capital surpluses due to historical accumulation (Smith, 1984, p125, p167; Harvey, 2004, p77;).

Secondly, capitalism strives to eliminate all natural geographic boundaries and relatively spatial boundaries, and "comes to represent itself in the form of a physical landscape, namely 'an artificial space' created in its own image" (Harvey, 2001, p247). Old space is ceaselessly divided and reshaped, while new spaces in accordance with the intent of capital continue to be created. For instance, for achieving the high-speed flow of capitals, commodities, money and labor forces in space, capital has been invested in the spatially fixed and immobile physical infrastructure of transport and communications systems (ports, airports, transport system, telecommunication) (Harvey, 1989, p182-183; 2003, p63) [4,5].

Capital being invested and fixed in some places will lead to the differences in productivity and the possibility of generating surpluses (Smith, 1984, p99; Harvey, 2004), because of the relativities of labor, location, investment environment,

---

4   As Harvey (2001) argued, "the annihilation of space through time" and "the annihilation of time through space" are considered as the only way to resolve the excess capital and labor, and the consequent economic crisis of capitalist mode of production, based on Marx's theory of capitalist crisis. They refer respectively to 1) reducing production costs and increasing surpluses by minimizing the turnover time of capital and speed up production, marketing and consumption; and 2) reducing the friction of distance by virtue of territorial divisions and specializations of labor).

5   Uneven development refers not simply to the geography of capitalism but also to uneven rates of growth (Smith, 1984, p99).

infrastructure, as well as other factors of production caused by uneven investment. That is, the "'homogeneous' global space is produced into the differentiated 'absolute space'" under the impetus of capital investment (Smith, 1984, p117). However, Smith also points out that "a tendency towards the equalization of the conditions in every sphere of production, and of the level of development of the productive forces is the nature of capital" (Smith, 1984, p114). Since the uneven geographical and spatial structure—the so-called "superimposed result of capital accumulations" of Smith (1984, 1990), capital is always attracted to flow towards the areas that are characterized by low-cost and high-profit to eliminate the spatial differences.

Third, based on the points above, Smith (1984) proposes a concept, namely the "See-saw movement of capital". In accordance with his exposition, capital tends always to flow from a developed area to an underdeveloped area where capital is more readily profitable. Capital flows bring about "the development of areas with a high rate of profit and the underdevelopment of those areas where a low rate of profit pertains" (Smith, 1984, p197). Thus, after a "downward flow", the underdeveloped area turns into a relatively developed area with low rate of profit compared to before (Smith, 1984, p198-). Meanwhile, the original developed area turns back to the underdeveloped area with relatively higher rate of profit. Capital movement on the global surface is guided by the rate of profit (Smith, 1984, p197), and the mobility further leads to the changing differentiation of geographical space.

As a kind of unstable results, Harvey (2001, 2004) concludes that 1) new geographical spaces continue to emerge, while the old spaces are periodically sacrificed; 2) the formation of differentiated geographical landscapes has become an important driving force of contemporary capitalist mode of production; 3) and likewise, the differentiated geographical landscapes in multi-scale geographical space also demonstrate the accumulation of capital in history.

2.3.2 Extracting value from cities

Lefebvre (1991), by constructing a concept of "abstract space", illustrates that there exists a conceived space in the world compared to the material space in reality (absolute space). In his opinion, this abstract space is the space of the power of the bourgeoisie and of capitalism, which resembles a lens and offers a specific orientation from which to view the world (Lefebvre, 1991, p51, 57). Following this idea, neoliberalism in contemporary practices, which praises highly of capital liberalization and financialization above and below the national scale, will inevitably give rise to such an abstract space. At the international and national levels, the space reflects a "unitary world-system with a division of labor and multiple political and cultural systems" (Jessop, 2012, p93), and the power of currency and national politics (Chang, 1999) which relies heavily on a vast network of financial institutes, commercial centers and main production centers (Kotz 2015, p13). At regional and urban levels, it indicates a kind of capital-oriented development and efficiency-oriented management patterns. It is summed up by Brenner and Theodore (2002, p368-372) as specific measures, involving the privatization and marketization of urban housing, public sectors, and labor market; large-scale projects of central business districts, industrial parks, and upscale communities; and the entrepreneurial governance and the "growth-first" strategy.

With the penetration of capital at the local level, cities—as the most concentrated areas of population, businesses and information, the cultural birthplaces of spatial subjects, and the institutional laboratories for policy and strategic experiments—have become the main venues for capital investment, and therefore are undergoing a significant transformation. Capital structures urban space as well as the political and cultural life associated with it (Arnould and Thompson, 2005, p22). "City itself is an architectural form, but also represents 'a microcosm of the abstract spaces', which is fashioned, shaped and invested by social activities during a definite historical period" (Lefebvre, 1974, p73). Urban (re)development and the consequent upgrade of the physical landscape are the most significant footprints that capital flows remain in space. After capital involvement and the interactions of related socio-economic activities, a steady stream of new "man-made landscapes" starts to emerge. It ultimately brings about a unique, dynamic physical and socio-economic network, which is constituted by both "hardware" (e.g. buildings, transportation, communications, functional zones and boundaries) and "software" (e.g. information and money flows, and socio-economic relations) (Harvey, 2004; 2001).

Subordinated to dictates of capital, cities have been gradually influenced by commercialism. Soja (1989, p97) points out that contemporary urbanization and the development of the urban environment are "a process of collective consumption." This collective consumption, as Castells (1979) defined it, by its very nature is the manipulation of the built environment, the extraction of urban rents, and the setting of land values, the organization of urban space, as well as the (re)production of goods, services, and labor. In furtherance of these processes, capital plays a vital role through investment, purchases, and/or intervention, and in which capital simultaneously achieves its appreciation and accumulation.

However, the engagement of capital in urbanization is more dependent on speculation (Lefebvre, 2003, p159) by leveraging the rise of real estate values as well as spatial investment strategies to facilitate (re)valuation of real estate properties and the surrounding built environment (Goldman, 2011; see Shin and Kim, 2015, p4). Capitalists are so willing to bind themselves to the real estate sector since it serves "as the best 'buffer' in the event of economic depression" (Lefebvre, 2003, p159), as capital is able to freely flow in and out and simultaneously to appreciate, or at least to hedge against inflation.

Real estate is a kind of commodity embedded in a particular location (Weber, 2002, p521). Because of the specific locational attribute, it is characterized as "scarce and valuable" (Weber, 2002, p521), and has always attracted investors in the community, from individuals to large investment institutions. By virtue of a clear ownership and the possibility of free transactions, property owners as well as investors are enabled to "capture any socially produced increases in ground rent plus the value of the improvements" (Weber, 2002, p521). However, owing to the spatial fixity, real estate is easy to devalue (Harvey, 2001; see Weber, 2002, p521). Thus, in order to hedge against inflation, continuously investing in the update of the functions and appearances of buildings as well as the physical environment or in the redevelopment of the original site of land is very important for investors and property owners. It is because that the constant updates and redevelopment will bring about potentially high rents and investment opportunities as well as surplus values.

Schumpeter (1934) attributes this constant renewal to the nature of capital—that is, the restless search for increased profits by means of making way for the new and devaluing the old simultaneously. Frequent urban demolition and reconstruction accompanied by large (re)development projects have become the prevalent approach for capital in contemporary society to recalibrate values, and what is more, to open up new spheres for further profit-making and accumulation. Today, such speculation becomes rampant because neoliberalism gives a good alibi for the iterative reconstruction—i.e. obsolescence and innovation (Shin, 2015, p4; Weber, 2002, p532). By removing barriers to the unfettered in- and out flows of financial capital in both global and domestic settings, capital is allowed to be fixed in the form of spatialized capital at any time. It can also be withdrawn in the same manner. Thus, capital circulates through the built environment in a dynamic and erratic fashion (Weber, 2002, p521). Along with the circulation, the build environment is purposely dismantled and selectively reconstructed. That is to say, the newly created spaces are placed high hopes on generating surplus value, while after a period of time they are bound to suffer destruction for the purpose of a new round of circulation. Thus, as we have seen, the physical appearances of old industrial facilities, and community structures as well as urban fabrics are being eliminated in many countries and regions nowadays; at the same time, plenty of CBDs, high-tech industrial parks and luxury condominiums are popping up everywhere. In this process, the real estate sector, characterized by speculation, high-profits return and non-permanent fixing, plays a crucial role in the production of "artificial spaces," creating surplus value in the secondary circuit of capital (Lefebvre, 2003, p159; Shin, 2015, p4; Miro, 2011, p2). Real estate and financial sectors come to play a very prominent role in assisting in the expansion of capital. Land as well as the ground attachments, similar to other scarce resources, serves as the main sphere of creating surplus value through real estate intervention and private investment.

2.4 New urban governance in neoliberal era

2.4.1 Cities in the neoliberal era—global, national and local interactions

As Peck and Tickell (2002) noted, neoliberalization has spread a new regulatory regime across space, which involves national, regional and local scales. However, some observers believe that t "neoliberal market rule is occurring with particular intensity at the urban scale and especially within re-configured urban policy regimes" (Picton, 2009, p47). Drawing on a synthesis of empirical and theoretical evidence, many neo-Marxist geographers and state theorists, represented by Bob Jessop, David Harvey, Neil Smith, Neil Brenner, Jamie Peck and Adam Tickell, have been engaged in fierce debates in the last several decades over the urban transition in the context of neoliberal shift, as well as the positive and negative impacts of neoliberal shift on local economy and society. Their topics are closely around the following aspects: capital accumulation by dispossession caused by capital liberalization and state's deregulation (Harvey, 2004; Castree, 2008); privatization of urban public space; local governance transition and neoliberal localization (Peck and Tickell, 2002; Brenner and Theodore, 2002); the shift from the Keynesian welfare state to the Schumpeterian workfare post-national regime (Jessop, 2000, 2002); the weakening regulation and social exclusion in urban space (Gough, Eisenschitz and McCulloch, 2006), and so on. These studies have offered a macro-analytical framework to reveal why neoliberal ideology and its accompanying institutional apparatus associated with global financialization have promoted urban elites to make changes in local regimes as well as governance practices, and how these ultimately bring cities with diverse configurations and historical-economic-cultural backgrounds to coincidently follow on a neoliberal path of governance?

Despite the controversy is ongoing, there seems to be a widely accepted consensus. That is, the analysis of local transition and urban policies should be closely linked to their supra-local economic environment, to international regimes and supranational blocs/institutions, and to the logic of serving global capital accumulation. Following this main line, state theorists and regulation theorists emphasize an analysis of the incentives of a universal shift from managerial (welfare-based) governance towards the entrepreneur (growth-driven) model through an in-depth observation of contemporary global production patterns.

Globalization and new challenges to cities
As described in Chapter 2.2, the wide spread of neoliberalism since the 1980s is closely associated with the rise of financial capital in the world, and the consequent requirements to a geographical integration of finance markets and the more flexible pattern of capital accumulation. However, the neoliberal restructuring and the rise of financial markets at both global and local levels have made regions/cities fall into a highly uncertain geo-economic environment, no matter whether they are voluntary or reluctant. The increasing economic uncertainties are characterized by "monetary chaos, speculative movements of financial capital, global location strategies by major transnational corporations, and rapidly intensifying inter-locality competition and so on" (Brenner

and Theodore, 2002, p367). This has brought both opportunities and challenges to cities/regions, and thereby leads to the rethinking and readjustment of national and local states regarding both the form and content of their political and economic governance.

The liberalization and deregulation of financial capital have facilitated the vast majority countries to be incorporated in a global capitalist economy, because of the frequent economic interactions among transnational corporations, financial institutions as well as wealth individuals (Hall and Hubbard, 1996, p159; Leitner and Sheppard, 1998, p292). The extensive economic exchanges on a global scale, characterized by an accelerated mobility of investment, have promoted rapid flows of capital between countries, regions as well as cities. On the one hand, global financialization makes it easy for cities/regions to attract capital from elsewhere (Leitner and Sheppard, 1998, p292), accompanied by the cross-national/regional investments. It thereby brings potential opportunities for cities/regions to become the economic and innovative leader of the world, and to rise up to the top of the global urban hierarchy. On the other hand, it is also likely to make some places lost the favor of capital, or even make them encounter the risk of the outflow of funds. As such, this rapid, flexible capital mobility will eventually lead to new urban orders, where jobs and investments move quickly, from city to city, up and down the urban hierarchy (Short and Kim, 1998, p56).

To safeguard local prosperity and to further promote economic growth, cities/regions have to engage themselves in "a fiercer international battle, fighting with one another over the attractiveness of capital investment and market", rather than relying merely on local firms and money (Leitner and Sheppard,

1998, p292). Given the deepening "global-local disorder", characterized by disorder, rapid change and uncertainty of the contemporary economic landscape (Brenner and Theodore, 2002, p367), and the arising potential opportunities, almost few cities can escape from but being involved in the competitive games. In order to occupy more shares in market (capital, resource and labor), the original urban regimes have been challenged, and new forms of governance with strategies are put forward by urban elites (Hall and Hubbard, 1996, p159). This repositioning, as Hall and Hubbard (1996, p159) and Soja (1989, p172-173) argued, aims to excavate or create comparative advantages to compete with their opponents in the international "wars for jobs and dollars".

State-local rescaling and local entrepreneurial transition

However recently, scholars have argued that the shift towards an entrepreneurial mode of operating cities/regions should not be totally attributed to the so-called "passive response" to the trend of global integration. In their opinion, states, regions and cities are not the "helpless pawns" of international capital (Parkinson, Foley and Judd, 1988, p2; Hall and Hubbard, 1996, p159), waiting for mobile capital being withdrawn from them). Quite the opposite, they are more likely to be a kind of independent entities who are fully capable of determining, or at least planning their own destinies (Hall and Hubbard, 1996, p159; Parkinson, Foley and Judd, 1988, p2).

As mentioned before, globalization and localization often occur simultaneously, and they both function as a conflation of homogenizing and localizing influence in spaces. Confronted with international and domestic environment changes, political agents (especially the state) play an essential role in serving capital accumulation by formulating strategies, and meanwhile in maintaining their hegemony through contesting power (Gramsci, 1971). Hence, examining state's decentralization is very necessary since it provides an alternative perspective (a response at national level to the global trends) to illustrate the motivation of local state transition. Decentralization of both autonomy and responsibilities from national to subnational levels has led to a rescaling of central-local relations (Leitner and Sheppard, 1998, p294). On the one hand, the reduction of unified arrangements from the central state means that regions/cities are not able to enjoy additional perks (e.g. preferential policies and fiscal subsidies to reduce regional inequalities) from the state anymore. Thus the most direct consequences are that localities are increasingly confronted with potential risks of fiscal shortages and unequal opportunities, as well as economic uncertainties caused by such a tremendous shift in international and domestic political and economic environment. On the other hand, the decentralization also leaves growing responsibilities on localities for social stability and economic prosperity.

Owing to the dual pressures from the withdrawal of central support on finances and policies and the enhancing responsibilities for local, cities/regions turn gradually into "cash-starved" entities that are desperately eager for growth. They re-set their targets to maximize local-based interests, to facilitate the favorable conditions in which profitable capital accumulation is possible (Hubbard and Hall, 1998, p14), to fabricate tales of municipal turnaround and urban renaissance to create new economic opportunities (Jessop, 1998, p91), and to enhance competitiveness to attract global capitals, labors and resources (Paddison, 1993, p340; Jessop, 1998, p82, p91; Leitner and Sheppard, 1998, p296).

2.4.2 New governance practices in entrepreneurial cities

Given the external economic uncertainties and internal growth pressures, more and more cities/regions have been induced to jump on the "bandwagon" of interurban competition (Peck and Tickel, 2002, p393). However, some of them are involved in the "place war" voluntarily, while others are reluctant. It is largely dependent on their locational characteristics, and the situation within the broader political-economic system (Haider, 1992; Leitner and Sheppard, 1998, p295). As Leitner and Sheppard argued that,

> "for some cities/regions, the external and internal changes provide them with new opportunities for prosperity through competitions and a means of overcoming unsuccessful development paths and economic stagnation; for others, however, it has undermined the reliability and success of past development practices, increasing the pressure to compete in order to reinvigorate the urban economy and retain economic prosperity (Leitner and Sheppard, 1998, p 295-296).

However, by no means a coincidence, both of them have invariably chosen an entrepreneurial direction. Cities/regions compete to throw themselves into economic, political and social innovations, to enhance local productivity and relatively stable competitiveness (Jessop, 1998, p82; Jessop and Sum, 2000, p2289). It helps to strengthen a stable and permanent capacity of localities to attract the footloose capital, residents, and investors at the expense of other places (see Jessop and Sum, 2000; Wu, Xu, Yeh, 2007, p19);

rather than securing the temporarily comparative advantages merely through tax breaks, subsidy cuts and deregulation.

Jessop and Sum (2000, p2289) stress that the "innovative strategies or 'new combinations' are characterized by real and reflexive". It is deemed as the by-product of the rethinking of local elites on the increasingly turbulent environment in both global and local. This reflexivity emphasizes on "the devising and realization of new ways of doing things to capture and stimulate development under particular local conditions (Harvey, 1989, p15), and more importantly, to generate above-average profits (e.g. the relative surplus value in Marxist terminology) in the course of capitalist competition (Jessop, 1998, p83; Schumpeter, 1934, p76, p154; McCann, 2004). To make a further illustration, Jessop (1998, p84-85) summarizes the contents of this innovation in two aspects, i.e. institutions and structures that directly support the existing entrepreneurs, and that help to sustain an entrepreneurial climate. Based on Schumpeter's works (1934, p129-135), Jessop and Sum (2000, p 2289-2290) summary it as follows:

1. Introducing new types of urban space (e.g. multicultural cities and cross-border regional hubs) so as to trigger new economic activities relating to living, working, producing, servicing, consuming, etc.;

2. Creating location-specific advantages by inventing new methods of space production, involving the updates of physical and technology infrastructure—the "hardware configuration", and the policy and program innovations relating to socio-economic environment—the "software configuration" (Jessop, 1998, p84; Belle, 2015, p8);

3. Building new markets in cities, and creating a new demand for the "space consuming", through modifying the "spatial division of consumption" (e.g. gentrification and enhancing living conditions), and "marketing specific cities in new areas" (e.g. festivals and cultural events, and promoting mega-projects (McCann, 2004);

4. Opening up new sources of supply to strengthen competitiveness of cities, for example by exploring new patterns of labor and investments, finding new sources of funding, reskilling the workforce, and introducing advanced technologies;

5. Redefining city position in global urban hierarchy (e.g. setting goals for upgrading as a world-class city), and refiguring the forms of organization in the global production networks (e.g. promoting inter-regional cooperation).

Besides, Jessop and Sum (2000, p2289) supplement that the innovative strategies are formed upon the basis of the self-reflexivity of promoters (or political elites) who praise highly "entrepreneurial spirits" on the real changing environment as well as dynamic crisis. Thus, there are some other issues being worthy of further exploration: namely who are the main actors who play roles in promoting the entrepreneurial strategies, who benefit from it, and what is the process of interactions between the promoters and opponents?

Strategic assemblages of actors: promoters, alliance-based mechanism, and entrepreneurial strategies

Some scholars standing from the perspective of political economy theories, such as Peterson (1981, p21), have argued that the growth of city is compelled by external market forces, and it is undoubtedly to the "benefit of all residents". And others suggest that the strategies are promoted by a group of politically mobilized local elites who are simultaneously the main beneficiaries of this shift (Logan and Molotch, 1987, p69; Cox and Mair, 1988; Clarke and Gaile, 1998, p13). In following with this logic, business interests of "entrepreneurs" should be considered as the most fundamental driving force to promote local economic growth. In contemporary capitalism, the business interests are increasingly fixed to the land in the form of ground attachments—i.e. the "local dependence of capital" as Cox and Mair (1988) and Hall and Hubbard (1996, p160) argued, and the "fixed capital in space" of David Harvey (1989, 2001). The fortunes of these "place entrepreneurs" have been tied closely to the rising land and property rents, accompanied by the economic growth at local (Short and Kim, 1998, p57). Based on these, it is not difficult to understand why "entrepreneurs" spare no effort to promote local prosperity.

As Jessop argued, the entrepreneurs might be local dignitaries and business elites, might also involve a wide range of actors, coming from official agencies, semi-official organizations, business associations and trade unions, citizens' communities, voluntary organizations, educational, religious and cultural institutions, and so forth (Jessop and Sum, 2000, p2291). These actors are not isolated from each other, playing their respective roles in local development. Instead, they are always organized into a kind of interest-based (business) coalition (Short and Kim, 1998, p58; see also Peck and Tickell, 1995). Besides, these actors are mobilized, "through a collective project" which is conductive to realizing their respective private gains, and "through a group of institutional factors that helps to consolidate their support" (Jessop and Sum, 2000; p2291).

In the neoliberal era, a pursuit of local interest growth—including economic prosperity and employment growth—has been placed on the top priority, and been "established as political phenomena and subsequently institutionalized" (Peck and Tickell, 1995). In this sense, business elites, by means of their powerful financial strength, are given an unusually strong "power of speech" in urban governance and formulating local political and economic strategies. Compared with other social groups, they are easy to forge alliances with local administrators as well as political elites, who grasp the legitimate authority in local management and are, in parallel, eager for an external funding of urban construction. Thus, as we have seen in many places, the state has alliance with private capital, and local political and business elites reach various partnerships on urban affairs. In this context, their efforts are more inclined to launch projects and development strategies in favor of capital accumulation, and to formulate liability- and profit-allocation schemes that are centered on the interest of business elites.

However, the capacity of business and political elites to consolidate and/or to expand their partnerships, and the strategies through which they achieve their goals, largely depend on the institutional embedding (involving institutional settings, interpersonal and inter-organizational relations, institutional constraints, etc.), as well as on broader economic, political and socio-cultural factors. Strategic Relational Approach (SRA), developed by Jessop (2001, p1223–1226), has establish a dialectical relationship between macro-economic-political structure and the selective strategies within particular contexts. According to this theory, local strategic choices depend largely on the dominant actors and their preferences. Thus, just as we have seen, in different socio-institutional locations and at different geographical scales, the way of entrepreneurial mobilization and the corresponding strategic choices are slightly different. For instance, urban entrepreneurial approach operated in the US and the UK is more dependent on the spontaneous social and market forces, while in the cities of East Asia, its implementation relies heavily on the excursive power of the developmental states (see also Belle, 2012).

Nevertheless, compared to their discrepancies, the commonality of the entrepreneurial innovations on strategies and institutions among distinct places appeared to be more significant. It is simply a byproduct of global neoliberal restructuring combined with local governance characteristics. Jessop (2002, p467-468) claims that the entrepreneurial innovation is just a "renewal and reinforcement of neoliberal principles" at urban dimension. The trend towards privatization and deregulation in economic spheres, and the attendant liberalization in social and political spheres, in fact imply a strategic choice of the dominant socio-political forces at national level. Correspondingly, a shift towards entrepreneurial governance at local—including local state transition and the consequent policies aiming to stimulate local economic growth—can be largely attributed to the institutionally regulated and policed disciplining of the state. Thus, in a certain sense, entrepreneurial urbanism serves as a local consequence of the "hegemonic project"—as neo-Gramscian scholars defined—of some kind of superstructures.

Owing the "top-down" penetration from national level, actors at urban dimension have unconsciously fall into the game that is deliberately designed by the state. To serve for, or cater to the neoliberal economic strategies that are selectively adopted by the state with a purpose of serving capital accumulation, governments at lower levels shift their eyes away from welfare provision, while more and more focus on the policy innovation that in favor of free market competition as well as flexible flows of capital. Besides, they also launch supporting projects as a testing ground for these policies, with the aim to attract capital at local and cultivate multi-partnerships. By analyzing the intersections between New Economic Policy (NEP), New Urban Policy (NUP) and Urban Development Projects (UDPs) in a neoliberal setting, Swyngedouw, et al. (2002) have revealed the links between neoliberal ideology and its associated economic hypotheses, and the application of its governance practices in the form of new urban policies and development projects (Figure 2.4).

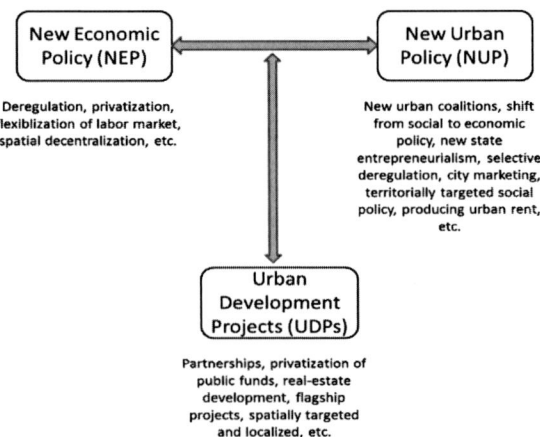

Source: Swyngedouw, Moulaert and Rodriguez (2002, p548)

Figure 2. 4 Relationships between NEP, NUP and UDPs

84

Specifically, the practices of local governance with a neoliberal sign can be noticed in the following aspects: (1) the doctrine of urban management shifts from managerialism towards entrepreneurialism—that is, abandoning the "top-down" Keynesian intervention, and pursuing a growth-first approach to local development (Peck and Tickell, 2002, p394; He and Wu, 2009, p282); (2) local authorities change their role from the suppliers of collective consumption to the defenders of favorable environment of global free competition, and meanwhile, the focus of policy measures shifts from social redistribution to promoting conditions favorable to competitive markets (Wilson, 2004, p771; Peck and Tickell, 2002, p394; Hubbard and Hall, 1998, p14); (3) the competitive advantages and economic growth rather than pure welfarist goals as well as local business regulations (Short and Kim, 1998, p58) have become the most crucial factor to evaluate a successful (or unsuccessful) governance (Hall and Hubbard, 1998); (4) public sectors are outsourced to non-state and quasi-state bodies, capital forces step onto the main stage of urban governance, and a variety of public-private cooperation are proactively extended (Leitner and Sheppard, 1998, p297; Hall and Hubbard, 1996, p155); (5) private properties, especially the land ownership is the fundamental basis of actors to conduct governance at local, and to establish cooperative relations as well as exercise civil rights.

The microphysics of governance—drawing on the concept of Foucault's "governmentality"

By drawing on a political-economic approach and the concepts of state and regulation theories, the macro-structuring principles that facilitate accumulation regimes have been given sufficient attention—in particular, with respect to the relation between economic regimes, local practices and the "extra-economic" conditions (e.g. Swyngedouw et al.'s model). However, this analytical framework seems to fall short in identifying the microscopic mechanisms to probe the process of practices and confrontations that (re)shape central-local and state-societal relations (MacLeod, 2001, p 822). It ignores the procedure of power exercising functions through some particular strategies. Simultaneously, it also neglects the potential for local and society to confront the hegemony and the strength of their resistances.

On the one hand, a single perspective from state and regulation theories gives an impoverished illustration of how hegemonic projects are implemented at the local level through deliberate policy-design and strategic innovation of the state[6]. Further, it ignores a truth when local elites embrace/refuse the top-down national strategies, they will mobilize the engagement of actors by introducing new urban policies in compliance with central blueprints, or intentionally mislead them. (Uitermark, 2005; MacLeod and Goodwin, 1999; Harding, 1997). On the other hand, this analytical framework ignores the micro-practices of resistance and cooperation in the local setting. It underestimates the importance of the interactions among local stakeholders and the possible and subtle effects on central and local decision-making (Künkel and Mayer, 2012; Uitermark, 2005). Hence, an analysis of the multi-scale origins of urban policies through the assemblage of theoretic approaches has been highly advocated by many scholars

---

6   In this study, the state is considered to be equal to "a place for class struggle and social forces" by drawing on the view of neo-Marxist scholars (e.g. Poulantzas).

in the last decade (Uitermark, 2005, p138). A neo-Foucauldian perspective on the analysis of neoliberalism and its influence on local governance currently fill this gap, by examining the enactment process of both the central and local political rationalities through governmental technologies (MacKinnon, 2000; Uitermark, 2005).

Governmentality, according to Foucault, is defined as the basis for political thought and action that relates to a specific form of power trying to dominate all of other types of power. More precisely, it refers to "an ensemble formed by the institutions, procedures, analysis and reflections, the calculations and tactics that allow the exercise of this very specific albeit complex form of power" (Foucault, 1977, p62). When people discuss neoliberal strategies within the context of governmentality, the discourses of the hegemony of the market over the state and the withdrawal of the state become techniques of the government (Lemke, 2001). Thus, for Foucault (2009), the dissemination of neoliberalism and its application in urban policies, in fact, mark a fundamental change in the techniques for governing subjects of political power with what he defined as a subtle, more delicate way (i.e. a suite of empowering technologies). This is different from the traditional forms of governance such as sovereignty and discipline.

From the neo-Foucauldian perspective, neoliberalism is understood as a form of trans-sectoral rationality (economic, political, social, etc.) based on the naturalization of entrepreneurial values, competitiveness and competition, and the individual accountability of entrepreneurs themselves (Morange, 2011). Following this logic, the main purpose (and also toughest task) of the governing power in the neoliberal context is to explore how to mobilize the self-governing capacities of individuals or groups in line with the principles of a competitive marketplace and the strategies or techniques conducive to the transformation of subjectivity from powerlessness to active entities.

To illustrate this point, Foucault (2009) has introduced two important concepts, namely "political rationalities" and "technologies of government," both of which are connected to the well-designed programs of government enacted for the purpose of taming individuals or groups with entrepreneurial spirits and, simultaneously, guiding their acts to correspond. First, he dissects a kind of intellectual machinery of political power, which is used for rendering the world (or reality) thinkable in such a way that it is always amenable to its political aspirations (Rose and Miller, 1992). This intellectual machinery is constituted by knowledge, languages and precise calculations—through which neoliberalism is no longer just an ideological project but is translated into a kind of noble spirit such as the "autonomy", "self-realization", and "self-esteem" of subjectivity. Its aim is to create self-regulating and self-responsible subjectivities in contemporary capitalist society by training their economically rationalized behaviors, which are determined by calculating the costs and benefits of the possible conducts.

Second, he points out that political rationalities must be underpinned by the operation of programs through special strategies, techniques and procedures or, as Miller and Rose (1990, p183) stated, "technologies of government" (Raco and Imrie, 2000, p2190). Such technologies are related to a complex assemblage of diverse forces—i.e. legal, architectural, professional, administrative, financial, and judgmental aspects—in relation to the decisions and actions of individuals, groups, organizations and populations (Rose and Miller, 1992, p281). They are exploited by some political forces to render programs operable and to make these entities understand the programs as well as strategies, thereby regulating themselves in accordance with the authoritative criteria.

These political rationalities and government technologies appear in both the national and urban dimensions along with the transfers of power and knowledge between the governing and governed actors. This is intuitively reflected in the relations between national and local actors and also between the governing (local central forces) and governed actors (vulnerable groups) at the urban level. At the national level, the economic crisis that happened in the 1970s was first problematized along with the identification of economic recession, stagflation and rising unemployment, and this was consequently attributed to the policy failures of Keynesianism. For instance, Thatcher's government criticized the power of the labor union and high-welfare policies and interpreted the crisis of capitalism as a crisis of governance. As a remedy, a set of neoliberal policies and relevant programs (i.e. fiscal cuts, local autonomy and other initiatives related to flexibility and austerity) were proposed because it is believed that they would "work better and deliver economic growth, productivity, low inflation, full employment or the like"(Miller and Rose, 1990, p4). Neoliberal policy is, thus, rendered as an effective solution to the perceived issues that are problematized by central authorities at the local level. Meanwhile, the paradigm of good governance, which aims to establish the optimal conditions for market-driven growth, is highly praised. This process has not only crystallized the active role of the local state as an entrepreneur (Leitner, 1990) but also established new institutional forms and "flanking mechanisms," which are intended to sustain and safeguard the entrenched accumulation strategies and interests of powerful economic actors and places (Brenner and Theodore, 2002; Jessop, 2002).

Similarly, local actors also have the ability to develop their rationalities and technologies in relation to the environment in which they operate (and may seek to manipulate) (Uitermark, 2005, p152). After receiving the top-down "instructions," central actors at the local level (e.g. political and business elites) compete to adopt new governance means—following with the principles of efficiency, flexibility, competitiveness, selective deregulation, collaboration and city marketing—to mobilize other actors to chase a common goal of building "entrepreneurialized" cities (Harvey, 1989, p4; Hall and Hubbard, 1998, p8-9; Wilson, 2004, p771; Picton, 2009, p47). Transferring the responsibility for urban construction to the public and innovating public-private cooperation is two of the main strategies of the neoliberal government. This shift at the local level implies that the whole society has embraced the market as a social and economic regulator. The "market morality" is successfully popularized in shaping the conducts of individuals/groups and accordingly assisting the capital in capturing an indispensable position during the process of urban development (Miller and Rose, 1990; McCann, 2004).

Although Foucault has strongly emphasized the power as the dominant role, the activities of the governed actors (e.g. civil society) are in fact not only an execution of or a resistance to the strategies and policies of authorities, but they are also constitutive of such strategies that contribute to some potential forces behind the evolution of governance (Uitermark, 2005, p144). On the one hand, the prospects, concepts, ideas and strategies of local actors (regarding the new form of governance and urban development) are governed by the power on the central scale. Besides, the ways of construction and cooperation at the local level are largely subjected to/in compliance with the aspiration of the central actors at the local level. On the other hand, the strategies of central actors are likely to be affected by the pressure that is put on them "from below" (e.g. negotiations of local stakeholders). Additionally, national actors are driven by the pressure they feel "from aside" (e.g. local negative cooperation and confrontation to the central state) (Uitermark, 2005, p152).

Chapter 3 Institutional Transition in China since 1980s: Authoritarianism Encountering with Neoliberal Discourses

3.1 Two voices on China's neoliberal shift

In the context of globalization, neoliberal doctrine has an important impact on economic theory and practices in many countries, and China is no exception. The political and economic practices in the last three decades show that China's economy is moving along a neoliberal trajectory (Wu, 2010) characterized by its launch of sweeping economic reforms. The state withdraws itself from many fields of social welfare provision; meanwhile, marketization and monetization are widely applied to then system of urban land, and housing as well as other public resources, such as the reforms related to state-owned enterprise restructuring, pension socialization, and the establishment of energy market (water, electricity, coal, refined oil, natural gas, etc.).

However, some scholars believe that China's reform during the last three decades was different from the practices of some developed capitalist states over the same period. That these two seemingly identical initiatives as well as their similar effects originated from different institutional regimes is just a coincidence (Liu, 2008). Others argue that the Chinese neoliberal shift is essentially an incomplete introduction of (or a selective reference to) the economic measures that were first proposed by Great Britain and the United States (Yuan, 2006). It is an experimental imitation of the reform of the Soviet Union as well as some Latin American countries in the 1990s (Vogel, 2011; Shirk 2007). Yet, how far the concept of neoliberalism can be (or has already been) applied to the

Chinese regime is debatable (Luo, 2012) since the neoliberal concept is proposed on the basis of the institutional framework of a capitalist regime. At the present day, China is neither a pure communist nor pure capitalist regime. Its re-orientation from a "planned economy" to a "Socialist Market Economy" has occurred only in the last 40 years. The reform is still at the exploratory stage, and there are too many uncertainties. For now, the neoliberal steering seems to be limited only to some economic fields, while this ideology is rarely applied in terms of politics, culture, ethics, and so on (Wu, 2010; Zhang, 2013; He and Wu, 2009).

On this point, some scholars hold different perspectives. They argue that the so-called neoliberal ideology today has germinated in China (Harvey, 2005; Wang, 2008, p14 and p98; Wu, 2005). The reform and opening-up policy adopted in the late 1970s is regarded as one of the milestone events of neoliberalism's expansion (Zhang, 2013, p3). China has made significant progress toward a market-oriented economy since 1978 (Bian and Logan, 1996), partially through the privatization steering of social materials and production organization (Harvey, 2005) and the phenomenal change in social welfare provision (Duckett, 2004). This shift is highly consistent with the turn to neoliberal solutions in Britain and the United States. That is to say, these two different regimes share a common ideological core — that the elevation of the market takes precedence over all other forms of organization (Mudge, 2008).

First, both regimes praise highly the "efficiency supreme" and emphasize that price mechanisms and market regulation are a valid approach to maximize socio-economic efficiency. According to this principle, it is understood that governments sometimes would not hesitate to use all political powers to serve the capital market. Second, neoliberalism advocates a complete (or moderate) reduction of state intervention. The role of the state is limited to creating and preserving an institutional framework for capital circulation and unfettered competition, while beyond these tasks, "[the] state's intervention should be kept [to a] minimum" (Harvey, 2005). Under China's regime, it seems this is the main trend of its reform in recent years. For instance, the country has started to admit the decisive role of the market in the national economy, reduced state intervention as much as possible while retaining necessary administrative regulations, cut down the controlling interests that governments hold in state-owned enterprises (SOEs), prohibited the direct involvement of politicians in business and management activities, and so on.

3.2 Neoliberalism with Chinese characteristics

Although there are several similarities in the practices of the reform of these two regimes, their socio-economic systems are essentially different. Brenner and Theodore (2002) have pointed out that it is rather necessary to distinguish neoliberalism as an ideology and a practice when discussing neoliberal practices at the local level. It is also essential to understand the "Chinese paradigm" of neoliberalization. There is no unified paradigm of neoliberalism in the world, and the ideology is often combined with local cultural and social-political backgrounds, and then produces a series of new measures as the implementation effects. Thus, the most prevalent template of neoliberalism and the existing experiences in many countries (e.g. Britain, the United States, Latin America, and Russia)—the so-called destruction and discretization of "Keynesian-welfarism and social-collective institutions"—cannot fully explain the Chinese characteristics (Ong, 2007; Zhang and Ong, 2008; Hoffman, et al., 2006; Wu, 2010). On the contrary, Chinese paradigm as a kind of new mutation, or as "a derivative of global neoliberalism", should be analyzed by linking the ideology with its social and political contexts.

To this end, David Harvey gives an accurate definition of the Chinese paradigm in his book "A Brief History of Neoliberalism". He refers to it as a "neoliberalization with Chinese characteristics" — through paraphrasing Deng Xiaoping's concept of "socialism with Chinese characteristics (or privatization with Chinese characteristics) (Harvey, 2005). Unlike some typical neoliberal states (e.g. European and American countries), China's neoliberal transition is rather unique, the essence of which is the combination of neoliberalized economic management and authoritarian state forms (Peck and Tickell, 2002, p37; Zhang, 2012). On the one hand, the free market mechanism, deregulation, and privatization have been vigorously applied in many social-economic areas since 1978. Besides appropriately relaxing the state's interventions in business practices and financial organization to mobilize the enthusiasms of capital, individuals and localities are executed. On the other hand, egalitarianism and the absolute leadership of the Communist Party over national politics and economy are still seen as a stable and long-term goal (Zhang, 2013; Wu, 2016). The state, both as the "destroyer" and market builder, has created a modern state out of its totalitarian and parental role in the past (Wu, 2010), while at the same time holding sufficient "discourse rights" on adjusting the breadth and the degree of neoliberalism in social and economic operation and management. Thus, deregulation and reregulation have alternately occupied the center stage (Wang, 2008), accompanied by the fierce struggles and compromises between the radicals and conservatives, and between different interest groups in society, and/or distinct factions of the ruling party.

The rise of a Chinese paradigm — the combination of indigenous ideologies with adventitious ideas

Wang (2008, p90) argues that China's reform is a forward orientation derived from various ideological interleaving and the constant (re)corrections upon its repeated practices. Thus, radical market doctrine, conservatism and new authoritarianism are entangled closely with each other, and jointly constitute the guiding philosophy of China's neoliberalism (Wang, 2008; Huntington and Liu, 1994). Zhang (2013) argues that China's gradual transition towards a neoliberal orientation is an outcome that the party-state conducted "a self-regulation" in response to the domestic economic depression and the deactivated society in the 1970s and 1980s.

However, when China carried out its reform and opening up policy, it was also the occasion of neoliberalism being in the ascendant in the West. Hence, its neoliberal experiment should be neither seen as a sudden turn of domestic ideologies, nor a whim of political elites in seeking changes after encountering setbacks in the planned economy period. Instead, it indicates the subtle influences of the global ideology of contemporary capitalism on China's traditional ideologies and existing institutional regimes. More specifically, it is a process that political elites made small concessions to the society that has been filled with resentments and discontents, so as to rebuild the legitimacy of its ruling. Furthermore, it relates to the fact that, political elites seek cooperation with domestic and international capitals at the expense of transferring partial benefits to ride out serious economic difficulties.

From the outset, China's market-oriented reform has not been deliberately designed along the trajectory of neoliberalism. Rather, it was just generated from a profound rethinking on the overly-strict command economy regimes in the Maoist era. Thus, the market-oriented reform in the early stage was hesitant and the initiatives at the time were relatively conservative. For example, in the previous National People's Congresses, the Chinese central government continually reminded itself and the whole society that political and economic reform should avoid an indiscriminate imitation of capitalist, and it meanwhile took pains to emphasize the legitimacy of the Communist Party of China (CPC) in ruling and socialism as the official ideology.

With deepening reform, as the whole country has tasted the sweetness of its initial economic reform, both radicals and conservatives of the CPC came to realize that the residual concepts and production mode under the planned economic system shackled the hands and feet of the country's further economic progress. For instance, throughout the 1970s and 1980s, the most intractable problems that the country faced were the low productive efficiency resulting from the system of state-owned property, and rising prices and corruption because of the "dual track" pricing mechanism. Led by the reformers represented by Zhang Ziyang and Deng Xiaoping, these issues that had long untouched were finally onto the reform agenda, after the reformists and conservatives had reached a short-lived agreement. However, their debates have never stopped, and the balance of their forces in affecting the reform has gone through changes. In this regard, it is primarily embodied in the "forward and backward" steering of the reform, such as the repression of the student movement in the 1980s, the privatization of housing, social welfare cuts, financialization, the establishment of special economic zones since the late 1990s, the enhanced control of the media control, and the remodeling of socialist ideology in the 20th century.

Obviously, China's reform has been accompanied by a self-adjustment of the party (Zhang, 2013) with restless contests between its interest groups, the self-adaptation of the party-state to the expanding capital markets, and the arousal of civil society in view of a changing domestic socio-economic environment. Nonetheless, the discourses of neoliberalism and globalization accompanied by frequent global trades and economic exchanges are increasingly powerful in affecting the orientation of China's reform. On the one side, Chinese political elites, especially the reformers are so urgent to integrate the country into the world and share the economic "cake" with the global. After some neoliberal ideas were introduced by the reformers in the mid-1990s[1], so-called Chinese mainstream economics were born. The economic and institutional bases of totalitarianism that had once prevailed in the country has been replaced by a so-called "market economy" which incorporates both neoliberal elements and is interdigitated with authoritarian centralized control, has established. On the other side,

---

1   Deng Xiaoping's famous speech about keeping on socialism or capitalism: "It doesn't matter whether a cat is white or black, as long as it catches mice". And the reform of free market and decentralization that were first practiced in the rural in the 1990s.

when China was approved to access to the WTO in 2002, it was forced to accept "Globalization-Plus" drafted by the US—i.e. China must promise to open its doors and accept the international investments, and moreover its national laws and regulations must be consistent with WTO rules. China's commitment to liberalization has far exceeded that of other developing countries (e.g. India) when they got permission to access the WTO. As such, the strong discourses and ideology of neoliberalism gradually penetrate in many spheres of national policies, and the strategies and practices of subsequent reforms as well as the value of social orientation.

However, owing to the continuous struggles among domestic and international ideologies (e.g. new authoritarianism, residual totalitarianism and neoliberalism), the debates between conservatives and radicals, and between the political power and capital forces within the country are ongoing. Additionally, because of the strong voice of opposition of the conservatives within the party, and also because of being subjected to the limit of its own ideology and foresight, the choices of the reformers with regard to reform initiatives are circumscribed. Thus, essentially, neoliberalism has never completely served as the mainstream in contemporary China, but is constantly substituted by other ideologies and then conversely penetrates others. In practice, China's reform indicates several particularities compared to other socialist countries and while showing sophisticated contradictions. At first, its reform during the past decades is moving along a more gentle, peaceful and gradual path, compared to the drastic transition ("shock therapy") of 1990s' Eastern Europe. Second, with regard to the reform areas, Chinese political elites that advocate economic reform should be prior to the political reform. That is to say, the party-sate endows the national economy with a great freedom, while limiting the spread of liberalization in the political, social and cultural fields. Besides, they consider that the prerequisites of marketization and democratization are necessary to establishe a highly centralized system of government and to create political entities that hold strong and effective decision-making abilities. All along, a tacit understanding has been reached between the conservatives and the reformers that all reform decisions must be carried out within the framework of the CPC's authoritarian regime.

This has brought about an unequal-triangular-relationship constituted by the state (power), market (capital) and society (civil). Even though confronted with pressures from economic growth and the awakening of civil society, the central state sometimes has to delegate partial authority to the local level, the market, and society. Precisely because of the ceaseless struggles and compromises of indigenous ideologies with adventitious ideas and among different interest groups (central and local state, market, and society), China's neoliberal practices inevitably exhibit a few mutable and paradoxical features. For example, contradictions exist between the voices about further decentralization from local and the existing strong central interventions on local development; between the gradual deepening market-oriented transition and the strong wills of authorities to regulate markets in a turbulent situation; and between the liberalization commitments to international investments as well as markets and the state's protection of the domestic economy amid the wave of globalization.

3.3 Market-oriented reforms: Reshaping the relationship between state and market

3.3.1 Introducing market mechanism and rebuilding the state-market relation

How to deal with the relationship between the state and market is a thorny issue that all market economies have to confront. Obviously, it is also the one that China has been deeply troubled about and committed to solving since the late 1970s. On the one hand, inspired by the "complete competition and free market" principles—that market mechanism is the most effective tool to allocate resources, and excessive government's intervention will reduce economic efficiency, the Chinese central government was rather urgent in seeking a feasible reform. On the other, however, the ruling party has been worried that the excessive and radical market-oriented reforms would probably undermine the legitimacy of its administration. Thus, in order to encourage society to accept the idea that a self-regulating market should succumb to state regulation, the Chinese central government uses an eclectic interpretation and expresses reservations on the market as well as the market mechanism (i.e. "the invisible hand) according to the national circumstance:

> Excessive government intervention would induce corruptions and economic inefficiency, and a 'reasonable marketization' will not impair social fairness and justice, on the contrary it could bring about higher productive efficiency and economic effectiveness (The Decision of the CPC Central Committee on the Establishment of a Socialist Market System);

Or rather:

> It is necessary to recognize the validity of the market, but also to be vigilant against the defects of the market (e.g. inequitable distribution of wealth, financial capital monopoly, periodic fluctuations in the economy and other issues); simultaneously, it is necessary to conduct proper government interventions, while avoiding excessive interference (The 3rd session of the 14th meeting of CPC National Congress).

Even though, the state has openly and officially acknowledged the importance of the market in the national economy, China has in fact been undergoing a long and bumpy exploration of the practices on how adjust the relationship between the state and the market. In the last four decades of reform, the attitude of the state towards the market mechanism in the allocation of public resources has undergone a subtle and gradual shift. The state's ambivalence can be easily perceived in its ambiguous definition of the market and in its slow and hesitant initiatives in the previous CPC's National Congresses (CNCs). With regards to the status of market in national economy and productive organizations, central government defined it as an "auxiliary role" in 1982, and then re-defined it as a "basic role" in 2002, and eventually emphasized it as a "decisive role" in 2013. Correspondingly, the proportion of state-owned property of public resources has been declining, and administrative interference in the allocation of resources has gradually weakened. The final result is that the state eventually shifts its function as a regulator or/and a supervisor at the macro level by discarding its "obsolete role" as an "all-round" entity in every sphere of the national economy and social security (table 3.1).

Step 1: plays an "auxiliary role" (1982-1989)

As early as the 12th CPC's National Congress (1982), the discussion on reshaping the relationship between the state and market was first proposed in view of the economic slowdown, the shortage in the supply of commodities, high unemployment, and so on that had emerged since the 1960s. In this congress, the CPC emphasized that the self-regulation mechanism of a market discipline was necessary and beneficial to China's planned economy. Any Chinese economic system reform must adhere to a principle that the central state's plan should always play a decisive role in the allocation of resources. Based on this, the market mechanism would be tentatively applied to the process of production and circulation of commodities. Although market adjustment played an auxiliary and supporting role, it implied that the entrenched planned economic system would begin to loosen. Its significance lies in that the market mechanism was recognized by the CPC for the first time.

Immediately the discussion on the role of market regulation was further extended. A term for the market economy with Chinese characteristics—the so-called "Socialist Commodity Economy" which was the original version of "Socialist Market Economy"— was proposed by CPC in 1987 (in the13th CPC's National Congress). In this congress, the "auxiliary role" of the market was further clarified. Simultaneously, the combination of state planning and market regulation in the national economy was legally put forward. The regulation of the market mechanism was allowed to be practiced in the production and exchange in a few fields, such as agricultural and sideline products, daily commodities products, and service and repair sectors. It also provided that the role of the market is to guide the economic activities of partial enterprises (mainly the non-public ownership business), while regulating that all market behavior must be unconditionally subject to government supervision.

Table 3. 1 Re-definition of the role of market in the previous CPC's National Congresses

Date and Conferences	Definitions of the market
1982-1989 (the 12th and 13th CNC)	"To introduce the market mechanism in the planned economy", "to correctly divide the content and scope of mandatory plans, guidance planning and market regulation"
1992-1997 (the 14th and 15th CNC)	"To make sure that the market force plays an essential role in the allocation of resources under the state' s macroeconomic control"
2002 (the 16th CNC)	"To give a fuller play to the basic role of the market in the allocation of resources and build up a unified , open , competitive and orderly modern market system"
2007 (the 17th CNC)	"To introduce institutions to give better play to the basic role of market forces in allocating resources"
2012 (the 18th CNC)	"To leverage to a greater extent and in a wider scope the basic role of the market in allocating resources"
2013 (the 3rd session of the 18th CNC)	"To let the market decide the allocation of resources"

Source: collected from the Xinhua News Agency and sorted out by the author.

In this period, the most meaningful explorations of marketization were the introduction of the price mechanism into urban land allocation in 1982 and the tentative allowance for the transactions of urban land (with some certain properties, such as commercial property) in Shenzhen, Shanghai and other pilot cities in 1987. Additionally, in the Constitutional Amendment promulgated by the central government in 1988, it was further stated that "enterprises, institutions and individuals are permitted to transfer the land use rights in accordance with the provisions of the law".

Step 2: plays a "basic role" (1992-2012)
Since 1992, Chinese reform reached a crossroads after Deng Xiaoping's famous Southern Speech in Guangzhou. The law of value and the market mechanism were widely recognized and accepted after China had tasted the sweetness of its initial market-oriented reform. In the 14th CPC's National Congress, the role of the market was further emphasized as a "basic tool" of China's socialist economy and the market mechanism was encouraged to be applied to more fields and in more depth degree.

During this period, the ownership reform of state-owned enterprises was an important attempt. First, central government empowered independent production and management rights to the majority state-owned enterprises, and it allowed some of them to transform into the joint-stock business. Then, the government liberalized restrictions on the private economies. For example, in the Constitutional Amendment adopted in 1999, the central government pointed out that the non-public economy would be an important component of the Socialist Market Economy. The second essential shift is the establishment of labor markets, which was advocated in the third session of the 14th meeting of the CPC's National Congress in 1993. Under this policy, labor was allowed to flow freely through the establishment of a hiring system (or a contract system) and a labor and social security system. The state and its units were encouraged to withdraw from the tasks of employment arrangements and the provision of social security. Third, to meet the requirements of accession to the WTO, China promoted an externally oriented economic system reform in 1994 and simultaneously expanded the spheres of foreign economies. In addition, the urban housing system reform (the abolition of welfare housing distribution system and establishing a commodity housing market), fiscal system reform (the separation of central and local taxes), financial reform (the disconnection of policy-oriented finance and commercial finance), foreign exchange system reform (the replacement of dual exchange rate system by a floating exchange rate), and so on. All show China's efforts to explore the market-oriented reform in many economic areas.

With the further deepening of market-oriented reforms, it is obvious that market efficiency has received unprecedented attention in the whole country and that the functions of government were largely weakened both in depth and breadth in the fields of national economy and social security. Yet, the basic role of the market was strictly defined within the scope of government regulations. Additionally, the basic role itself contains an ambiguous semantic, and there is a lack of a precise definition on the extent and depth to which deep the market forces could apply this role to the national economy. Under this regime, the administrative power still held strong rights, thereby justifiably being able to compete with the market for the right of resource allocation in the name of supervision and regulation. Hence, the debate over the role of state and market in the national economy never stopped during this period. The central government's attitude towards the range and depth of applying market mechanism always wavered. Consequently, administrative interventions never disappeared, but were only constrained to some extent and temporarily weakened, accompanied by the sometimes relaxed and sometimes tightened regulation of the central state. For instance, national policies about the regulation of the real estate sector have drastically changed since 2013. Those policies with strong administrative colors that curbing housing prices by setting limits on the quantity of individual house purchase and the average sale prices in each city, have been loosened. Instead, many more economic and financial measures such as the use of property tax and credit regulation as well as increasing the housing supply are repeatedly mentioned in official documents.

Step 3: plays a "decisive role" (since 2013)
The fuzzy definition of the state and market relationship has caused either excessive administrative intervention, or government malfeasance. This can be easily perceived in the real estate market, the stock market, and the financial and energy sectors since the mid-1990s. In view of this, the boundary between market and government was further clarified in the third session of the 18th CPC National Congress in 2013:

> The core issue of economic reform is to address the relationship between government and the market, so that the market can play a decisive role in the allocation of resources (The State Council's decision on several major issues of the reform, 2013)[2];

> Government should acts as a referee rather than an athlete. Government's intervention in the allocation of resources is in

---

2   Source from: Xihuan Net, website:
http://news.xinhuanet.com/english/china/2013-11/12/c_132882359.htm.

order to make up for deficiencies of market, instead of replacing the role of market, (Renmin luntan, 2015)[3].

A relatively detailed re-definition of government behavior aims to draw a more distinct line between the government and market. First, it stresses that the market should play a regulatory role by virtue of its autonomy, rather than being subject to governmental regulation and controls. Second, it highlights the importance of the laws of value, competition law, and the balance of supply and demand in the national economy. Moreover, it argues that the administrative intervention should withdraw from microscopic fields of economy as much as possible.

Conversely, governments are called upon to be committed to providing policies and legal guarantees for market operations, so as to create a fair competitive environment for market players. Moreover, they are required to offer basic services, infrastructure, and special goods to the public in some areas in which the market is often out of order and causes social inequity. Specifically, it provides that the government must conduct macro-regulation for avoiding market failures and economic slowdowns; protect the provision of public goods (e.g. basic medical care, compulsory education, and public housing) and necessary social welfares (e.g. minimum living allowance, pension, and unemployment insurance); loosen administrative examinations and approvals on rent control, land leasing and infrastructure construction; and

---

3   Source from: http://paper.people.com.cn/rmlt/html/2015-03/20/content_1550176.htm.

provide a legal and institutional framework for the market (e.g. clear property rights).

However, the explanation about the decisive role of the market that the Chinese central government provided is different from the definition under the capitalist system. According to the official explanation:

> the term of 'market determinism' neither means that governments will entirely relax controls on the market, nor implies that the market will be tightly limited under the administrative supervisions as what the central government defined between 1992 and 2012; instead, it is intended to inspire the governments' role of playing as an effective and just actor — extracted from the Decision of the CPC on Some Major Issues Concerning Comprehensively Deepening the Reform[4].

Admittedly, the capability of central government for conducting the surveillance and control of the national economy has been gradually weakening since the market-oriented reform of 1978. Local governments are urged to withdraw their hands from some part of the micro-economic areas as well. In parallel, the rights as well as part of the duties and obligations have been largely peeled away from both central and local administrations, and these have been transferred to the market and society at the legislative level.

However, this is not to say government's intervention has disappeared. Central and local governments still have a strong

---

4   Data from: http://www.china.org.cn/chinese/2014-01/16/content_31215162_4.htm.

willingness and the potential capacity to participate in economic development and the maintenance of stability in society by virtue of their administrative powers (e.g. in the accountability system — local governments are asked to be responsible for local affairs, to review urban planning, to allocate state-owned resources, and so on.) and the monopoly in some of public resources (e.g. the ownership of urban land resources, and the control power of primary land market). Thus, even though marketization has received an unprecedented attention from the whole community, and the market-oriented experience of being put into practice is much more extensive and profound than ever before, administrative intervention is simply weakened but has not vanished. In fact, government regulation starts to exert influences in a more indirect and subtle way.

3.3.2 The establishment of urban land market

As China is making its greatest efforts in re-adjusting the relationship between the state and the market, and in reconstituting its political and institutional configurations towards market economy, one of the main purposes is to create every possible condition under its existing framework for economic growth and capital accumulation. The most effective and convenient way to reach this goal is by conducting a large number of urban construction projects, and to encourage the involvement of private capital in urban development[5]. Commercialization as well as privatization appears to offer a possibility for a wide involvement of private capital. Among a series of reforms launched

---

5   This is considered as the experiences learned from Hong Kong's economic prosperity of the 1970s and 80s.

by the state in the past decades, land and housing reform undoubtedly plays a rather crucial role in China's neoliberal urbanization (He and Wu, 2009, p287).

Land system in the planned economy

When talking about China's land market-oriented initiatives beginning in the 1980s, it is necessary to trace these back to its previous land system changes. In contemporary history, China has been through three tremendous land reforms, namely the privatization of land (before 1949), the public-oriented reform (1949-1952) and the market-oriented reform under the socialist regime (since 1982). In the first period before 1949, the Communist Party of China conducted a land privatization reform, in which it confiscated lands from large landowners and then reallocated the lands to sharecroppers and landless peasants. This reform aimed to eliminate the feudal land system and establish the private land ownership of farmers.

Then in the early period of the planned economy period (between 1949 and 1952), the second land reform was initiated by CPC to solve the unfair distribution of social wealth. During this reform, the property and land of private enterprises were either acquired or confiscated by the state, and all private properties were subsequently converted into state-owned or collective assets. Until 1958, China's socialist transformation had been basically completed, and over 90% of urban land had been nationalized (Zhang, 1997; Ding, 2003). Since then, China's land market has not existed, and for a long period, land was only considered as a non-commercial productive factor. This is completely unlike what the value of land is in the complete market economy that overwhelmingly determines the enterprise's location choice.

In fact, land at the time had only "use value" but no "exchange value." Between 1954 and 1984, urban land was regarded as a kind of free resource that was able to be accessed only through administrative channels (IFTE, CASS and IPA, 1992; see Zhu, 2005). Behaviors related to land transactions, including rental and exchange without official authorization, were strictly forbidden and were considered illegal. Although the administrative allocation was the only way that land users were able to access land, once organizations or individuals obtained the lands, they were allowed to occupy the land nominally without any time limits, and without any charges. Under the planned economy regime, where economic entities were allowed to settle on and how many lands they were allocated depends primarily on their attributes and the types of products they produced and, most importantly, on their political ties with central and local governments. For instance, state-owned enterprises (SOEs) and institutions (or rather work unit—"*danwei*"), as the primary profit-making and producing machines of the country, were undoubtedly the major beneficiaries who benefited from national policies and land allocation. However, in exchange for the national investment and free allocated lands, these non-independent economic entities were required to hand over most of their profits. In fact, land revenues along with land value were totally hidden behind the overall profits that SOEs created (Ding, 2003).

Under this system, lands were over-supplied to consumers compared to their real needs. Free resources encourage users to over-consume without considering affordability (Zhu, 2005). Moreover, due to the lack of widespread competition among applicants and simultaneously failing to give land users an effective incentive to conduct the intensive development of allocated lands, this system led directly to a serious waste of urban land. It resulted in the emergence of numerous factory courtyards (*danwei dayuan*—"单位大院" which is normally composed of warehouses and factories and living areas) springing up all over the city. Many of them occupied substantial efficient and highly productive land in the central locations of the city.

Land system reform since 1982

In face of these issues, a clear-cut state ownership of urban land was established through the Constitution Act in 1982. Immediately, in 1998 it was stipulated in the Constitution Amendment that "land-use right must be separated from land ownership" and "the right to the use of land can be transferred" (National People's Congress, 1988: Article 2). Because of the reform, land transactions are permitted in the manner of trading land-use rights, and urban land once again accesses to its "commodity attributes". Nevertheless, central government is still the sole landowner who monopolizes tightly the ownership of urban land.

Second, the paid-use way replaced the original gratis allocation way in this reform. Land users are allowed to possess land use rights within a fixed period of time, and the vast majority of them are asked to turn over a certain amount of land use fees in the fixed period. With the rise of the land market, three different approaches regarding land transfer were proposed by the central government, which contains negotiation, bidding and auction. Administrative allocation (*huabo*, "划拨") has no longer served as the sole approach of applicants in accessing to urban land. Instead, market starts to play a primary role in land allocation. In addition to a few types of land use (e.g. the lands occupied by the state organs, military and some public institutions[6]), the majority of urban lands are supplied by the governments through a competitive way (i.e. bidding and auction).

Under this regime, the state remains acting as the owner of urban land, and absolutely controls over the land supply in the primary land market (*churang*, "土地出让") (Yeh and Wu, 1996; Hong and Zhang, 2012). Nevertheless, the real-estate developers, companies, institutions as well as other users who have already gained the land use rights from governments after paying for a certain land leasing are allowed to transfer the use rights to others through the secondary or lower markets (*zhuangrang*, "土地转让").

This reform, in some degree, reduces the central controlling on the land transaction process, and helps to lower the threshold for economic entities accessing the land market. It is primarily reflected in the establishment of the secondary or lower levels' land markets. Meanwhile, the paid-use way has effectively alleviated the phenomenon of the non-economic and non-intensive land use, caused by the gratis allocation system under the planned economic system. According to the land price mechanism, factories, warehouse as well as other land users who cannot afford high land premium or rentals (most of them are SOEs or work units established before 1978) have to move out of the city center; in contrast, real estate industries, financial services and retail industries with higher labor productivities replace them and occupy the city center. Besides, by introducing the public bidding

---

6   These lands are normally allocated to land users by governments, or are transferred to users after the users paying for a small amount of rents based on negotiations between the two sides of governments and applicants.

and auction, it also helps to avoid the irregular operations and administrative corruptions, which were rather common over the past.

However, government interventions are still strong in the form of administrative allocation and negotiation, which inevitably bring about the information asymmetries, price randomness, illegal discounts and corruptions. For example, sales through negotiations are non-transparent deals where land prices can vary very much (Zhu, 2005). This often leads to government officials in collusion with land users, so that the price discounts widely exist. Besides, the dual-track land system (i.e. the coexistence of gratis allocation and paid transfer) provides a breeding ground for the arbitrage of local officials and economic elites.

Another government's intervention in the market is reflected in the monopoly of the land supply—i.e. the government control over the primary land market (*churang*, "土地出让"). Owing to the lack of a perfect competition platform, governments as the sole providers of land are easy to manipulate the market by reducing the total land circulation, and thereby raising land price. Besides, governments have permissions to governance the secondary and lower levels land markets (*zhuanrang*, "土地转让"), namely the administrative power of performing approvals to all land transfers. In other words, free transactions among land users are, in fact, not allowed. The new purposes of land use and the trading prices are normally asked to go through the strict scrutiny of local governments. For example, governments can overrule some land transactions when they do not meet the public or government's interests, or the requirements of relevant planning. Furthermore, another common intervention is that governments set a benchmark land price in localities, which aims to avoid the over-low or over-high prices of land transactions.

Especially in 2007, the central government introduced a "Land Reserve and Management Act". In this Act, it is required that the lands which are going to be redeveloped, owing to the projects involving enterprise restructuring, urban renewal, industrial enterprises relocation, etc., or because of the promulgation of the new urban planning should be first purchased and stored by local governments. After gaining the land use rights, governments conduct some necessary infrastructure construction, and then transfer the use rights to the society through the market. After the enactment of the Act, land transactions in the secondary market are essentially equivalent to that in the primary land market.

3.3.3 Urban housing reform

As another important composition of China's market-oriented reform, the privatization and commercialization of housing reform are another significant steering towards the neoliberal orientation. Compared to the land system reform, housing system (supply and security) reform seems to be more comprehensive and thorough. It is characterized by a widespread, in depth privatization and commercialization reform within the housing supply system, and the greatest degree of deregulation of the housing market.

To begin with, it is necessary to give a brief introduction to China's welfare housing system between 1949 and 1980. Between 1949 and 1980, China adopted a socialist, work unit-dominated welfare system and a low-rent public housing system, which is similar to the practices adopted in most socialist countries. This system arrangement draws a mixed thought from three components:

socialist ideology, welfare philosophy and clan tradition (Zhao and Bourassa, 2003), with a principle of eliminating private ownership and diminishing social inequalities and class exploitations (Lee and Zhu, 2006, p41). To achieve this goal, the vast majority of private housing throughout the country was converted to public ownership by governments since 1956 (Wang and Murie, 1996; Chen, 1998). Meanwhile, in consideration of maintaining fair and just, housing construction was only allowed in a non-profit way after the 1956's public-oriented reform of urban housing. In other words, housing construction was deemed a low priority compared to other capital investments. Until 1978, the amount of public housing accounted for 74.8% of the national urban housing (Liu, 2011).

Under this system, public housing and workers' housing were built according to the social demands of the time, and all construction and allocation tasks were mainly borne by the work units as well as municipal housing management bureaus. First, municipal governments on behalf of the state collected implicit income taxes from workers in the form of low wages, in which the taxes were used to provide housing, education, medical care as well as other welfare subsidies for citizens. After obtaining the taxes, central state distributed a certain amount of taxes to municipal governments, local urban housing bureaus as well as work units for the purpose of public housing construction and maintenance. Work units and administrative departments then allocated the housing for employees and citizens by collecting a small amount of rent from tenants. Urban citizenship identities (i.e. age, seniority, position, education background, and household size) were considered as the only references to access to the public housing, in which the seniority, job position and education background are the most essential indicators to determine the quantity and area of the public houses that individuals obtained.

The urban housing system characterized by the low wage-implicit taxation and low rent (Chen, 1998, p 44), and that lack of investment of social capitals since municipal governments, urban housing bureaus and work units acted as the sole responsible institutions for collecting construction funds, have given rise to several negative effects. Because of the low rent, both governments and work units were unable to afford the fees of maintenance service for the existing public houses. From 1950 to 1980, the expenditure of housing rent was only 6% to 9% of the average household income in Beijing (Xie, 1999, p3). In sharp contrast, the actual funding needs for housing maintenance and management largely exceeded the affordability of both governments and work units. Thus, due to the lack of timely maintenance and redevelopment, urban housing conditions throughout the country were rather poor and shabby.

In addition, due to the insufficient funds, construction speed of urban public housing was unable to be consistent with the growth rate of demand for urban housing. Especially, in the context of the rapid influx of agricultural migrants to urban areas with China's fast industrialization, this contradiction was increasingly serious. According to records from the Information Database of China's Reform[7], there were about 86.9 million households being in urgent

---

7   Date is collected from the Information Database of China's Reform, website: http://www.reformdata.org/content/20100909/6711.html.

need of housing in 1978. This amount accounted for 47.5% of the country's total urban households. Besides, owing to the supply shortages, the per capita living space fell to 3.6 m² in 1978 from the amount of 4.5 m² in 1950. Until 1989, there were 469,481 households living in a rather poor condition that the per capita living space was lower than 2 m² (Chen, 1998, p44).

Another negative consequence arising from the public housing system is a potentially unfair distribution of public housing among different types (properties) of work units, and different individuals (Lee and Zhu, 2006, p41; Wang and Li, 2008,). Firstly, construction funds or housing resources of different work units accessing to, were unequal. This top-down allocation approach destined to that the work units at higher administrative levels were much easier and more likely to get funding supports or land and housing resources from central government. By virtue of their close political and/or economic ties with national and local administrations, they have much more abilities to bargain with the central state. Consequently, non-state-owned work units (primarily collective enterprises) and the communities of jurisdiction of urban housing bureaus were facing with more serious housing shortages than that state-owned work units and administrative units faced. At the end of 1990s, in the total number of work units confronted with difficulties on the supply of welfare housing, 74.4% were non-state-owned enterprises, while the rest was state-owned enterprises and administrative units (Yang and Wang, 1992).

Administrative allocation way and low-rent system (almost free) stimulated the desires to take up free resources as much as possible. It offered a breeding soil for corruption, and therefore, led to an unequal housing allocation. Individuals with strong administrative powers, such as cadres, tend to be able to get relatively abundant housing resources and those with a large area and a good quantity. Moreover, their privileges allowed them to help other persons gain the houses with larger areas and better quantity. The quasi-clan system caused significant inequity amongst different work units and within work units since 1970s and became an urgent problem that the government was forced to act (Lee and Zhu, 2006, p41).

In addition to the above points, the welfare distribution system also brought the work units a heavy burden in non-productive areas. Housing together with education, medical care, and pension as well as other social benefits were closely bound up with jobs and as a commitment in the work agreements which were signed between employers and employees. Hence, work units had no choice, but to spend a large amount of labors and financial resources to support their commitments each year, even when they encountered financial difficulties. Moreover, the commitments and agreements forced the labor force to be greatly immobilized in one place (Lee and Zhu, 2006, p41). Simultaneously, workers were inclined to stay because the length of service in their work units was normally a very essential indicator which would be considered preferentially in the allocation of welfare housing (Lee and Zhu, 2006, p41; Chen, 1998, p44).

From this, China's housing reform was urgently needed (Wang and Murie 1996). In order to prevent the failure experiences of governance and to solve the resulting housing shortages and inequalities, the old unit-based and public housing system was broken. Meanwhile, the privatization and commercialization were introduced in China in succession since the end of 1980s. The responsibilities of raising funds, investment and construction that were shouldered by the central state, local governments and work units during the past are gradually transferred to individuals, financial institutions and property developers respectively.

After clarifying the housing property and establishing a housing market, in which the private property rights are admitted to being traded in the market, the welfare allocation is repealed. And individuals are encouraged to purchase houses according to their financial strengths. Subsequently, in order to ensure the housing purchasing ability of individuals, Housing Provident Funds System (HPFS)[8] and the banking lending system are launched respectively since 1998. These two systems aim to offer capital support for individual purchases by ways of compulsory savings (work units deduct part of monthly wages for individual purchase) and providing preferential financial loans. Besides, work units are asked to provide monetary subsidies to their workers in the form of increasing salaries, rather than providing physical subsidies— welfare housing over the past.

Stage I: commercialization and socialization of urban housing, the initial privatization, and improving residents' willingness of purchase (1980-1987)
Reviewing the overall process, China's housing system reform is sporadic rather than general (Lee and Zhu, 2006, p 42). Specifically, this slow and tortuous process consists of three phases. In the first phase (1980-1987), the reform aimed at introducing commercialization and socialization in the urban housing system,

---

8   Social Security Administration (SSA) is responsible for managing Housing Provident Funds System.

and simultaneously to test the willingness and abilities of urban residents' housing purchases.

The specific practices are to raise the rent of welfare housing, to sell partial public housing to individuals by charging a one-time transfer fee, and to ask government and danwei to increase subsidies to employees and residences (i.e. governments, danwei and individuals bear 33% of the housing purchase costs respectively). In this stage, the initiatives above were implemented and tested in several selected cities (i.e. Zhengzhou, Changzhou, Siping, Shashi, Yantai, Tangshan, Bengbu). Nevertheless, this reform is an important step in the privatization of public services under the socialist system. More importantly, it shakes the entrenched concept of socialist welfare housing and people's consumer attitude (Lee and Zhu, 2006).

Stage II: the deepening reform of housing privatization, and the preliminary reform of housing finance (1988-1997)
Given the initial success, the state conducted the second step of housing system reform since 1988. Practices of this reform included raising the rent of public housing, selling public housing to individuals, and increasing monetary subsidies to employees. Through these initiatives, central government hoped to cut the proportion of public housing and encourage housing privatization by admitting private property rights in the above pilot cities. Furthermore, its purpose is to make an initial test for the future reform in a broader region of the country.

In 1994, the State Council issued an act named "Decision on deepening the reform of the urban housing system". It marks that China's urban housing reform has entered a phase of deep monetization and marketization. In this Act, the term of Housing Provident Funds System (HPFS) was first proposed whose intention was to ensure the purchasing abilities of individuals. It forces that both work units and employees (not including those in private sectors) should make monthly contributions to partial saving accounts for the purpose of employee's housing purchases. Second, it encouraged the sale of public housing to individuals being promoted in a wider geographic range over the country. Third, several state-owned banks were appointed by the state to provide a kind of "policy loans" to individuals whom are going to buy the public houses[9]. Besides, public houses were allowed to enter the market after five years since individuals had acquired housing property.

By the end of 1997, the proportion of owner-occupied housing in urban areas was up to around 50%, and in some regions, the proportion was even up to 60%[10]. With the privatization of property right and trading liberalization, urban housing market started to take shape. Yet, the market was still flawed and was not run in open and fully competitive environment, and the most housing transactions at that moment were the purchases and sales

---

9   The policy loans are those granted by national policy banks (e.g. the National Investment Bank). People's Bank of China determines the total size of annual lending, and simultaneously limits the forms of the loans. Besides, the funds can be only used for non-profit public projects, in which the majority is sponsored by local governments.

10   Data is collected from the website:
http://baike.baidu.com/view/467274.htm?fromtitle=%E4%BD%8F%E6%88%BF%E 6%94%B9%E9%9D%A9&fromid=3114316&type=syn.

of public houses. Consequently, a true sense of real estate market did not appear. On the one hand, the state and work-units played an intermediate role between market supply and state-subsidized consumption as usual (Wu, Xu, Yeh, 2007, p49). On the other hand, individuals were yet lack of motivations to purchase private houses through the market, since the low-rent of public housing was still strongly attractive to them in the existing welfare allocation system. Additionally, given the few demands of the public for private housing, and central government's tight control over the housing market, professional real estate sectors did not grow up quickly in the whole country.

Stage III: the abolition of welfare housing, nationwide reform, and establishing a financial security system in law (since 1998)
Since 1998, China's market-oriented reform of urban housing entered into the third phase. The objectives of the reform at this stage are meant to promote housing privatization, commercialization, monetization in the whole country, and to establish a fully competitive market. After the State Council issued the policy entitled Notice on Further Deepening the Urban Housing Reform and Accelerating the Construction of Private Housing" in 1998, a more "radical" attempt—that is seemingly contrary to the core idea of socialism, while more similar to the practices in Western capitalist countries—was implemented. According to Lee and Zhu's view (2006, p42), "owner-occupation, a popular form of tenure in Western industrial societies, have taken root in China".

In this round of reform, the in-kind allocation in the work unit-based supply system was thoroughly discarded since 1998. Services involving housing constructions, repairs and management were assigned to some socialized and specialized agencies. More and more public houses were converted into private housing by selling the property rights to individuals. In the field of housing finance, the central government promulgated a Housing Monetization Policy (HMP) and a Housing Credit Insurance System (HCIS). In HMP, it is pointed out that the funding supports from HPFS, the small-scale loans provided by the Chinese People's Bank, and the commercial loans will extend to a wider population.

In addition to direct cash subsidies from employers in lieu of in-kind housing subsidies, urban residents are able to access to more financial supports. The significant purposes of HMP and HCIS are to encourage individuals to purchase private houses, in accordance with their real demands, family savings, and housing provident funds of individuals. Simultaneously, it is to reduce the risks from individual loans of housing purchases. However, the underlying purpose is to ensure a strong purchasing power and reliable capital guarantee for the growth of real estate market.

After the slow and lengthy period of exploration, China's housing system has already met a high level of privatization and marketization than ever before. It is especially noteworthy since 2003 when the State Council issued an Act, titled "Promoting a sustained and healthy development of real estate market". In this Act, it is argued that:

> In the next few decades, China will shift its efforts from "promoting housing market by means of affordable housing provision" to "increasing the percentage of commercial housing in the market by encouraging the construction of commodity housing".

It is meant that governments will fully withdraw from the role of housing supply, and allow the vast majority of urban housing in the market be supplied in the form of commodity housing. By this approach, they are able to be freed from their past heavy administrative and financial burdens, in relation to the tasks of housing subsidies, management and maintenance. They only assume a small amount of tasks, such as the investment and construction tasks of affordable housing for rarely few groups of low-income residents.

Since then, the vitality of the housing market has been greatly activated. The roles of governments and work units as the main providers of urban housing are gradually replaced by the specialized real estate companies and institutions. Accordingly, the national and work unit-based housing supply system is replaced by a monetary-based supply system which is subject to market regulation. It ultimately established the status of real estate as a pillar industry of the national economy, and makes the commodity housing become the main ingredient on the market. By the end of 2010, the percentage of housing privatization in urban areas of China rises rapidly from 30% in 1994 up to 89%[11]. What tightly accompanied by the privatization is the booming of the housing market in all Chinese cities. In particular, between 2000 and 2006, national investment growth rate of real estate is excess 20%, except for the 19.8% growth rate in 2005. In 2010, the investment of national real estate development reached 4.6 trillion yuan, of which the commercial residential investment amounted to 3.4 trillion

---

11   Data is collected from the website: http://www.reformdata.org/special/656/.

Yuan[12]. Recalling the past few decades, China's market-oriented housing reform (or real estate development-oriented) has been successful in increasing the amount of urban housing, stimulating economic growth, and improving the urban living environment. However, owing to the over-reliance on the market and the neglect on the government's responsibilities to adequate supply of public housing, and social equity, it has triggered several conflicts, such as housing inequalities, urban poverty, social polarization and spatial segregation, etc.

3.4 Decentralization and reshaping the central-local relations

3.4.1 Fiscal redistribution and delegation of authority

Another neoliberal steering in China's political and economic reform is the rescaling of power relations between central and subnational governments (He and Wu, 2009; Lin and Zhang, 2015). In particular, relationships refer to the right to dispose of local fiscal revenues, the authority of managing local socio-economic affairs, and the duty of promoting local prosperity. The reshuffling of state power has owed its origins to the adjustments of the national tax structure that were respectively conducted in 1971 and 1994.

Acting as an omnipotent provider, the state, together with its agents—state-owned enterprises (SOEs) and local governments had dominated the entire economy and society in China between 1949 and 1978 (Zhu, 2005). At the time, the central state performed

---

12   Ibid.

not only as a socialist welfare provider for its citizens, but also as a manager and a supervisor of national economy whose most essential tasks are to organize industrial productions and to spur economic growth on a national scale. Conversely, subnational governments acted as executors for the central state's commands without any autonomy. On the one hand, they assisted the state in handling local financial management, and economic and public affairs. On the other, subnational governments received a fixed amount of funds from the central state for the local spending for the next year, after they had turned over all fiscal revenues of the last year. Under this highly centralized and "top-down" fiscal system under the socialist regime, 80% of national fiscal incomes were tightly held by the central government. It seriously suppressed the enthusiasm of local authorities in local infrastructure constructions and economic promotions. Except for obeying the central commands, it was not necessary for them to carry out any innovations in local governance and policy operation.

In 1971, a fiscal contract system (*cai zheng bao gan* "财政包干") was initiated to transfer more powers and responsibilities to localities. This reform aimed to arouse the enthusiasm of local authorities on economic investments, thereby driving domestic economic growth. This reform was mainly related to the re-adjustment of the national financial system, which was accomplished through a re-segmentation of fiscal revenues between the state and localities. Under this regime, the allowances from central to local were abolished. Instead, subnational governments were required to hand over a fixed sum of revenues to a higher level annually, in accordance with their negotiations every five years (Lin and Zhang, 2015). The surplus finances were allowed to be left at the local level if they were able to create surplus income. Meanwhile, the central state relaxed its control over local development affairs, decentralized the power of decision-making and shifted a great deal of responsibilities of social and economic developments to localities (He and Wu, 2009; Wu, 2010; Lin and Zhang, 2015). In cities, the maintenance and development of urban built environment that is used to be taken into account by the central state in its regular budget were shifted to the shoulders of municipalities (Wu, 2010; Lin and Zhang. 2015). Furthermore, the state delegated and entrusted the management rights of a large number of SOEs to subnational governments. The most direct consequence is that incomes of local SOEs contributed to the vast majority of local fiscal revenues at the time.

Owing to this fiscal policy system, the central state had successfully shifted its administrative burdens and financial pressures to the local level. In exchange, however, the state paid a high price that the central control over local fiscal resource was greatly weakened. Conversely, subnational governments were the biggest beneficiaries because their decision-making powers on local affairs and rights to freely dispose of local finances were largely enhanced after the fiscal decentralization (He and Wu, 2009; Fu, 2015; Lin, 2014). They were permitted to carry out investment plans and construction projects in line with local needs. They finally got rid the image of passive "command executors," and performed more like local "managers" with strong profit-seeking motives, who bundled their achievements tightly together with the profit increase of SOEs and local economic prosperity.

Due to the decentralization arising from the fiscal contract reform, central powers had been largely restricted while it gave birth to a wave of relatively independent local authorities. This inevitably led to the potential interest conflicts between central and local governments. As subnational governments were granted the right to claim the residual revenues, they, of course, expected to retain the revenues at local as much as possible, even at expense of the interests of the central state. Local protectionism sprang up all over the country. In order to evade the share of fiscal revenues that should have been submitted to the central state, irregularities—a serious underestimation of local fiscal revenues, an overestimation of budgets and provisions of tax reliefs for local businesses[13]— occurred frequently. Meanwhile, since local incomes that the central state and localities respectively obtained were divided in accordance with a fixed amount but not a ratio, the revenues handed over to the central state from localities did not increase promptly with local economic growth. It enormously undermined the interests of the central state. Between 1978 and 1996, the Chinese economy had grown more than fivefold but central revenues were essentially stable (Lin and Zhang, 2015, p2781).

In order to avoid local fiscal evasion and to increase central revenues, and simultaneously to not harm that the enthusiasm of subnational governments devoting themselves to local development, a Tax Sharing System (TSS) was introduced in 1994. It indicated that the Chinese fiscal system had converted from profit remittance to a taxation levy (Zhu, 2005). The main approach to this reform was to adopt a standardized tax rate to GDP and a stipulated central-local sharing ratio in all taxes (Lin and Zhang,

---

13   Subnational governments conducted illegal tax breaks for enterprises at first, and then they recovered the funds in the name of "apportion" and charges.

2015, p2782). National revenues are divided into three parts, namely central tax, local tax, and state-local shared tax (Table 3.2). The central government holds 75% of the value added tax, 60% of the income tax, 100% of the resource tax and 100% of the business tax (e.g. revenues from transportation and finance and insurance). The rest composed primarily by land-related revenues (i.e. land transfer tax, property tax, land appreciation tax, etc.) is left at the local level. As a result, central fiscal incomes increased substantially from an average of 27.5% between 1978 and 1993 up to 52.3% in 1994 (Figure 3.1).

Table 3. 2 China's Tax Categories

Central fixed tax	Local fixed tax	Central-local sharing tax
Customs duties, consumption tax, VAT revenues collected by customs, income taxes from central enterprises, banks and nonbank financial intermediaries	Business taxes (excluding those named above as central fixed incomes), income taxes and profit remittances of local enterprises, urban land use taxes, personal income taxes, the fixed asset investment orientation tax, urban construction and maintenance tax, real estate taxes, vehicle utilization tax, the stamp tax, animal slaughter tax, agricultural taxes, title tax, capital gains tax on land, state land sales revenues, resource taxes derived from land-based resources, and the securities trading tax	Only the VAT is shared, at the fixed rate of 75 % for the central government, and 25 % for local governments
The remitted profits, income taxes, business taxes, and urban construction and maintenance taxes of the railroad, bank headquarters and insurance companies		
Resource taxes on offshore oil extraction		

Source: Wong, 2000, p7.

Additionally, a bi-level administrative system of national tax revenues was set up throughout the country (Lin and Zhang, 2015). The State Administration of Taxation (SAT, "guoshuiju") is in charge of collecting the central taxes as well as central-local sharing taxes. Local Taxation Bureaus (LTBs, "dishuiju") are responsible for local taxes. After SAT collects the central-local sharing taxes, it will allocate part of them to localities in accordance with a fixed ratio. The "fiscal re-centralization" through the establishment of a bi-level administrative system and a clear tax classification effectively protect the share of central fiscal revenue (Fu, 2015) and assist the central state in regaining the authority in public finance. As a result, it greatly diminishes the share of national revenue that subnational governments could access to.

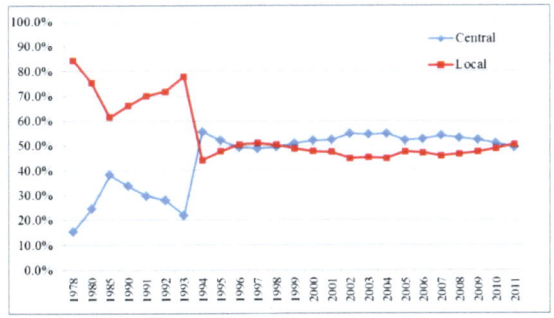

Source: collected from China Finance Yearbook and Tax Year Book of China.

Figure 3. 1 Proportion of central and local fiscal revenue between 1978 and 2011

However, the transformation toward a re-centralized economy is marked with a mixture of market mechanisms and planning controls (Zhu, 1999). It has not weakened the economic autonomy of subnational governments over local governance. Conversely, they still hold strong powers to organize economic production and manage local affairs, and moreover owe the strong abilities to exert influences on the allocation of monopolistic resources (e.g. urban land, energy, water) by means of tax regulation and to choose the approach of urban operation by virtue of their authority over urban planning and management. Under this regime, subnational governments are such kind of subjects who have special benefit structure and utility preferences and are able to achieve their administrative goals following their own wishes. Furthermore, they possess the potential strengths to bargain with the central state. In this way, it is sometimes difficult for the state to unilaterally impose its will upon the localities.

3.4.2 The "bottom-up" responsibility: a Cadre Appraisal System (CAS)

Fiscal redistribution and delegation of responsibility give rise to a "principal-agent" mode between central and local. In terms of economic and urban development, the central government has granted maximum autonomy to subnational governments (Lin, 2014; Lin and Zhang, 2015). However, it has never given up the supervision and control over its subordinates, and has been trying to set up a bottom-up "responsibility mechanism", or rather an up-down "accountability mechanism" to ensure that the subnational governments exercise the rights within the permissible range of the central (Lin, 2014; Wu, 2010; Edin, 2003). Thus, at the beginning of 1980s, the state introduced a Cadre Appraisal System (CAS).

The CAS is not only as an instrument of a higher level to control lower-level agents and to regulate central-local political relations, but also as a means designed to mobilize local initiative and to maximize administrative efficiency (Edin, 2003). Governments at higher levels carry out assessments to subordinates according to their performances (i.e. local economic growth, the increases of employment and local revenues, the attraction of inward investment, social stability, etc.), and then reward the outstanding ones or punish those do not complete the stated targets. Besides, central government hopes to build a competitive mechanism among local officials in the same hierarchical administrative level, in which the leading cadres are placed in an internal ranking order within their administrative regions on the basis of the evaluation results (Edin, 2003). This evaluation mechanism with a remarkable characteristic of encouraging competitions makes the political career prospect of local officials increasingly dependent on the accomplishments of economic development (Lin and Zhang, 2015, p2783) and local prosperity.

Through the fierce competition and the attractive rewards—i.e. the promotion of cadres, the administration and responsible relations between the central and local have changed. In most cases, central government and higher administrative departments will no longer directly issue the concrete instructions and commands on local economic, social and public affairs to the lower levels of government, just as the common practices in the planned economy period. Instead, the central state is more inclined to commission the duty of promoting local development to subnational governments, meanwhile specifies a minimum completion criteria (Wu, 2010; Wu, 2016). As a result, the "top-down" administrative arrangement under the planned economic system, in which the subnational governments normally played the role of absolute obedience to the central commands, is replaced by a "bottom-up" responsibility mechanism. Subnational governments use up all likely available resources, so as to assist them in achieving the goals, even better than their competitors at the same administrative level.

3.5 The arising land finance and urban management

Through the TSS reform, the state once again juggled revenue assignments yet left the expenditure assignments untouched (Wong, 2000, p15). The delegated responsibility and duty at the local level have not been reduced with the local financial cutbacks. Subnational governments are not only responsible for local profits and losses in urban capital operation, but also in charge of all social and public affairs (i.e. urban public affairs, infrastructure construction and the provision of social security services). Moreover, financial pressures from these responsibilities increase year by year, along with China's rapid urbanization and a steady stream of national infrastructure projects introduced since 2008[14]. Due to the sharp contradiction between the onerous local responsibilities on socio-economic management and the overstretched local budgets, subnational governments have fallen into an awkward situation. Since 1994, local fiscal expenditure has exceeded fiscal revenue, which results in a huge fiscal deficit (Wong, 2000; Lin and Zhang, 2015). At a rough estimate, subnational governments accounted for over 70% of total public expenditure while collecting less than 50% of total government

---

14   In 2008, the central government introduced an Economic Stimulus Program", named "4 trillion RMB (around 586 billion dollars) investment plan". It aims to ease the export setback because of the global financial crises erupted in the second half of 2008 by increasing the national investment in domestic infrastructure construction. After the introduction of the plan, subnational governments (mainly at municipal level) immediately invested around 18 trillion RMB in line with it.

revenues after the implementation of the TSS reform (World Bank, 2002; see Wang, 2014).

Against this background, local authorities attempt to explore all possible ways to generate extra-budgetary revenues, so as to resolve their fiscal dilemma. More importantly, they are keen to highlight their political achievements in localities according to the requirement of the CAS. However, there is a premise that subnational governments are not allowed to engage in business activities any more after the state-owned enterprise reform since the beginning of the 1990s[15]. Hence, tax revenues related to land development as well as the infrastructure construction that constitute the main part in local tax structure in the framework of TSS (Lin and Zhang, 2015) are certainly deemed by local officials as a "life-saving straw" to seize and as an essential guarantee for local revenue generation and their achievements.

Accordingly, an active pursuit of land (re)development and infrastructure (re)construction is seemingly a viable and indeed lucrative choice for local authorities. The legality of administrative monopoly on the land resource and government regulation of the land and housing markets, especially after the introduction of a dual-track land market since 1980s, have made it possible for them to engage in the commodification, upgrading, and expansion of urban land. These are considered to be an effective means to

---

15   In the Third Plenary Session of the "twelfth CPC National Congress" in 1984, the state-owned enterprise reform (SOER) was formally proposed. This reform requires clearing the line of ownership and management right of SOEs. Government officials are not allowed to engage directly in the activities of business and management of SOEs.

generate income, promote economic growth and promote cities' images.

When "land finance" (tudi caizheng) becomes a tool of making up for the financial shortfalls, the maximization of land value naturally turns into a common strategy in the majority of Chinese cities and towns. On the one side, subnational governments own the responsibility for the management and regulation of the land market. On the other, they are assigned by the central state to monopolize land supply in the primary land market (Lin, 2014; Lin and Zhang, 2015). In the situation of fiscal shortages, local authorities tend to expropriate agricultural land and some of the state-owned lands—those are the state-allocated lands in the planned economy period—at a rather cheap price. Afterwards, they sell the lands at a very high price, thereby reaping immense profits in the primary market ("churang" market). These profits are usually covered in the land transfer fees and land use taxes (Ding, 2003; Zhu, 2005). In addition to the huge price gap between the low-cost land acquisition and high-priced transfer of governments, land use tax and land value increment tax that are generated from land transactions between land users in the secondary market ("zhuanrang" market) are another important components in local land finance. Since 2000, the proportion of land revenues in total local fiscal revenue has increased substantially. According to China Land and Resource Almanac published in 1999, the total amounts of land conveyance fees were about 51.4 billion yuan, accounting for up to 9% of total sub-national budgetary revenue (558 billion yuan) (Ye and Wang, 2013). In 2011, however, the land conveyance fees amounted to 3.15 trillion yuan, which accounted for over 60% of total subnational budgetary revenue.

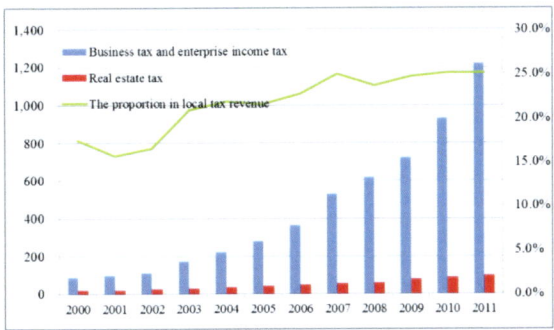

Source: data collected from China Finance Yearbook (2001-2012) and China Real Estate Statistics Yearbook (2012) and were calculated by the author.

Figure 3. 2 Tax revenues from real estate and construction sectors and its proportion[16]

According to the design of TSS, except for the taxes related to the land use, transaction, and development, those generated from the real estate and construction sectors (e.g. business tax, income tax and real estate tax) are totally owned by the local. Together with the land grant revenue, they constitute a major part of local fiscal incomes (Figure 3.2). This indicates that promoting real estate development and infrastructure upgrades has turned into an efficient manner to favor local revenue growth. It therefore, inspires municipalities to set their sights on the launch of new projects involving new town construction and inner city renewal (or urban regeneration). Since 2004, the proportion of real estate and construction-related taxes in local tax revenue in the whole

---

[16]   The indirect land taxes include business tax and enterprise income taxes of construction and real estate sectors, real estate tax, etc.

country has been maintained over 20%. In some individual cities, the proportion even rises to more than 30%. In Urumqi (an underdeveloped city in Northwest China), for instance, the real estate and construction-related taxes contributed 31.84% of local fiscal incomes in 2013, while this figure was only 8.97% in 2001[17].

Obviously, fiscal reforms cut the direct investments of the central state to localities; meanwhile market-oriented reform has largely weakened the direct control of governments over public resource allocation and management. Increasingly financial pressures and the resulting "land finance" have greatly distorted the behaviors of governments at various levels and their officials (Lin, 2014). In addition to the continual sale of land in exchange for the lucrative profit returns, another approach that attracting technological labor and capital investment by means of providing supporting services, favorable "hardware" environment and preferential policies is increasingly favored by local officials. They believe that the investment of land-related incomes in urban environment contributes to a "visible" political performance (Lin, 2014). Simultaneously, it helps to enhance local attractiveness and competitiveness to inward investors as well as inward capitals. It eventually brings about GDP growth and local socio-economic prosperity, and furthermore assists them in easing financial pressures and meeting the assessment requirements of the central state. As a result, an endless circle has emerged, which is linked by local revenue growth, governments' investment in urban construction, the influx of inward capital, creating of economic

---

[17] Data is from Urumqi statistical information website,
http://www.wlmqtj.gov.cn/stats_info/tjfx/20141124182621806.htm

growth and local prosperity, and further stimulation of the enthusiasm of local investment.

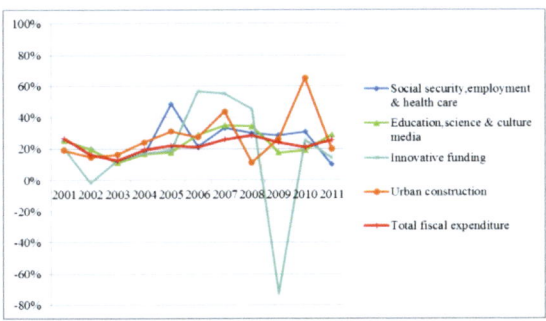

Source: data collected from China Finance Yearbook (2002-20012) and China Urban Construction Statistics Yearbook (2011) calculated by the author.

Figure 3. 3 Growth rate of local fiscal expenditure in public affairs[18]

Undoubtedly, a pursuit of land finance and attracting inward investments from international and domestic investors become an ambitious target, and also the most urgent task of local authorities. Through the analysis of local fiscal expenditure in the whole country, it can be found that the fiscal expenditures on urban construction since 2001 has always been ahead of other areas that local governments invest in, even though the rate was sometimes slightly lower than the overall level (Figure 3.3). And this is rather

---

18   The innovative funding refers to the fiscal expenditures in financial and service sectors.

significant during the period of 2008 to 2010, in which the state launched the "Economic Stimulus Program"[19].

In inner cities, governments pin their hopes on urban renewal and land reconstruction projects for the development of the real estate, financial, and service sectors. They invest directly in these, or more probably request the investment of social capital in infrastructure upgrades and the construction of large-scale commercial and public projects. Huge swaths of old towns are razed and rebuilt, replaced by a dizzying government buildings, theaters, convention centers, spacious and magnificent squares, and towering office buildings as well as luxury apartments. The new Central Business District (CBD) springing up in many cities, such as Shanghai's Pudong New District and Guangdong's Pearl River New Town. One of the main purposes of these large-scale construction projects is to attract domestic and foreign multinationals to settle down their headquarters there. In urban fringe and rural areas, they are eager to develop agricultural land into construction lands and to build new production places such as New Town or Industrial Development Zones in order to attract manufacturing enterprises. By the mid-1990s, the number of diverse industrial development zones had reached more than 6,000 all over the country, and its total area was over 15,000 km^2 (Yu and Zheng, 2003).

3.6 Summary and discussion

In this chapter, I have adopted the perspective of evolutionary game theory to examine China's institutional transition, arguing

---

19   Chinese government invested 4 trillion RMB in infrastructure construction to stimulate economic growth.

that neoliberalism and China's indigenous ideologies have been alternately dominant in the arena of public opinion regarding political reform. This phenomenon is accompanied by repeated negotiations among various interest groups, including reformers and conservatives among political elites, social organizations, groups, and citizens, as well as the economic elites newly emerging with the rise of domestic capital markets.

Here, I prefer to term China's institutional transition as a "neoliberal direction with a selective logic." On the one hand, the open and market-oriented path of China's reform was a strategic choice of the political elites when the entire country was wracked with severe economic recession, political instability, and social unrest in the 1970s and 1980s. On the other hand, this selectivity is obviously limited. It can be attributed, to a large extent, to the global neoliberal hegemony, with its powerful discourses and rules, at the local level and increasingly frequent trade and other economic exchanges between global and local markets. With the fierce contests between neoliberal discourses and indigenous ideologies, China's institutional transformation is full of twists and hesitations. One of its most significant features is an unsynchronized neoliberal shift in many spheres of the country. For instance, economic reform has far outpaced liberalization in the political, social, and cultural fields, leading to an unequal triangular relationship among state power, capital markets, and civil society.

The second feature is that state reform initiatives are sometimes open and aggressive, while sometimes conservative and stagnant. This is easily seen in the restructuring of state-market relations and central-local rescaling. Obviously, market-oriented reform has become the major objective of the state, which partially opened the door to international and domestic capital, allowed the private sector to participate in economic production, and encouraged urbanization and the rapid development of cities. Moreover, it gradually transferred some of the welfare state's duties to the market and civil society. However, governments still play an important role in resource supply and configuration, by virtue of their monopoly in ownership of resources like urban land, mineral resources, etc., and their authority to regulate and supervise markets. The role of the market has been strictly defined within the scope of government regulations, becoming a tool for governmental promotion of economic growth. In view of this, the range and depth of applying market mechanisms to resource configuration always wavered with the state's ambivalent attitudes and the varying tolerance of society towards liberalization and marketization. As a result, direct administrative intervention has actually been replaced by a subtle but powerful form of indirect influence. The state, through its selective use of the market, co-opts the power of capital to mobilize resources to achieve its political and economic visions.

The third feature of co-adaptation and integration of conflicting ideologies is the rescaling of central-local relations. Through the (re)adjustment of tax distribution structures and the reconfiguration of administrative power and duties in favor of local management, the once-dominant statism has disintegrated. Although decentralization appears to have been inexorably deepening in recent decades, the center has not forgone all interventions in local governance in its truest sense. On the contrary, the central state keeps a tight grasp on rescaling central-local relations by easing and tightening fiscal policies. Instead of the omnipotent role and ubiquitous influence it had in the past, the central state has adopted more subtle ways of exerting authority. For instance, it now assesses the successes and failures of local officials by introducing a Cadre Appraisal System.

Thus it can be said that both neoliberal discourses and policy paradigms are integrated in China's traditional ideology and planned economic regimes, all in the shadow of the state's targeted regulatory impulses. A new set of institutional arrangements has thus been generated, which has the characteristics of not only neoliberalism but also certain elements of a planned economy. This contradictory nature of the restructured institutional system can easily be seen. For example, it is embodied in the dual-track system in the land market, the state's residual yet still powerful impacts on the housing market, and the intense contests between central and local governments over finance and taxation. However, it is worth noting that, rather than locked in an eternally binary opposition, the contradictions are more likely to exist in a virtually interplay of mutual adaptation and integration. We have seen repeated adjustments of state-market and central-local relations over the last 40 years in China, rather than having one completely replace the other.

This hybrid institutional system plays the role of an intermediary that connects the global and the local, transferring information between them. It also sets new rules for the game, reconfigures some public resources by redefining property rights, and adjusts power and interest relations by rescaling both central-local and public-private relations. As a result, the contradiction and complexity are shifted to the urban dimension and the recalibrated rules have played a crucial role in repositioning the character of local governments, affecting the nature of target-setting, policy-making, and operating actions of local officials. On one side, the legitimacy of government intervention in the market and state ownership of property make it possible for governments to profit from markets, while on the other, the reshuffling of central-local relations leads to an extreme imbalance between local fiscal revenues and the obligations that local authorities have been asked to assume. Land financing has become the most important target of local officials; to maximize land value through relentless land (re)development and infrastructure investment has naturally become an effective and prevalent strategy among the vast majority of Chinese cities and towns.

Chapter 4 Urban Renewal Strategy, New Policies and Local State Transition

4.1 The rise of Lanzhou city

4.1.1 Becoming a national industrial base

Lanzhou is an ancient Chinese city on the Silk Road, situated on the upper reaches of the Yellow River. It is now the second largest city and an important industrial base in the Northwest China. Before 1950, the city was only a commercial and handicraft city. It had almost no modern manufacturing industries, but was represented by some manual workshops (Chang, 2002), and few light industrial enterprises built by the Qing government and the government of the Republic of China. According to the Lanzhou Economic History:

> Before 1949, there were just 36 enterprises (Lanzhou Power Plant, Gansu Coal and Mine Factory, Gansu Machinery factory, Lanzhou Automobile fitments Factory, Northwest woolen mill, Lanzhou Flour Mill, etc.) located in Lanzhou city, contributing the industrial output value about 15 million yuan... In the eyes of economic circles at that time, the city was still stuck in the Middle Ages...Up to 1949, urban population was less than 200,000, and the gross national product is only about 30 million yuan. (Wei, 1991).

After benefiting from a series of favorable economic policies as well as concentrated investment in large-scale industrial projects from the central government—the "First five Plan" (1953-1957), the "Second five Plan" (1958-1963) and the "Third Front Construction project" (1965-1971)—since the early 1950s, a modern industrial system consisting primarily of petrochemical and machinery-manufacturing industrial sectors was built in the city. It made the city one of the world's fastest growing cities in a new era of its industrialization after World War II (Wei, 1991). Thus, the rise and development of the city in the past decades has been in an inextricable relationship with its industrialization. Between 1953 and 1957 (the "First five Year" period), Lanzhou was included in the list of eight key construction cities (others are Wuhan, Xi'an, Chengdu, Chongqing, Hengyang, Shiyan, Baotou and Luoyang), and was planned by the state as one of three national petrochemical bases—the largest petrochemical industrial base in Western China. During that period, six petrochemical projects of the Soviet Union aided program (156 projects in total) were selected to settle in Lanzhou[1]. The petrochemical enterprises were led by Lanzhou's oil refinery and Lanzhou's petrochemical factory, and eight ancillary projects as well as several defense industrial factories were settled ten kilometers away from the city center and in the upper reaches of the Yellow River—the Xigu district, under the joint deliberations of China Ministry of Works and Buildings, Soviet urban planners, and Lanzhou City Construction Committee. To provide the petrochemical products for the whole country more conveniently, the state invested and built two railways lines in

---

1   The six Soviet aid projects (investment amount) are respectively Lanzhou Refinery Plant (182.23 million yuan), Xigu Power Plant (110 million yuan), Lanzhou Fertilizer Plant (251.8 million yuan), Lanzhou Rubber Factory (123.4 million yuan), Lanzhou Petroleum Machinery Factory (156.22 million yuan), Water Plant (37,16 million yuan) and Lanzhou Refining and Chemical Plant (72.81 million yuan). Data are from the Lanzhou Statistical Yearbook (1980).

Lanzhou at the time—in western direction connecting Urumqi city, and to east connecting Lianyungang city. In addition to the title of petrochemical base, Lanzhou acted as an essential transport and logistics hub at the same time, in contributing to national security and stability by connecting China's eastern and western regions.

Source: Lanzhou City Planning Bureau.

Figure 4. 1 Lanzhou City Master Planning (1954-1972)

By 1957, Lanzhou received a total of 720 million yuan of financial support from the state. In the same year, its GDP rose to 233 million yuan, with an annual growth rate of 48.5%. Even during the "Cultural Revolution" period, its economic growth rate was still maintained at about 13.2%. The scale introduction of national projects gave birth to the prototype of the heavy-based industrial system of the city, and significant economic achievements placed Lanzhou in an important position in China's politics and economy. In addition, it had a profound impact on the city's industrial structure as well as its consequent urban layout.

From 1958 to 1962, China ushered in the "Second-five year" period (or the so-called "Great Leap Forward" period). Inspired by the "successful experiences" of the Soviet Union, and propelled by the ambitions of competition with British and American industrialization, the central state set off a vigorous Iron and Steel Production Movement in nationwide. The development strategy in Lanzhou at the time, on the one side, concentrated on the further expansion of oil refining and chemical production lines. On the other, Lanzhou municipal government, with the assistance of the central state, set up a number of municipal-owned enterprises represented by iron and steel smelting industries (e.g. Lanzhou Aluminum Factory, Lanzhou Iron and Steel Factory, Lanzhou Thermal Power Factory as well as other steel production-related businesses) in response to the national strategy. There were a total of 102 industrial projects built or renewed during this period. By 1960, the industrial output value of the city had reached 1.52 billion yuan, while the indicator was 34.49 million in 1952, and in the terms of industrial output, heavy industry contributed about 70% of total gross industrial production.

Obviously, this movement further facilitated Lanzhou's industrialization process, and assisted the city in successfully squeezing into the ranks of newly industrialized cities. However, the attentions of governments at all levels were locked on the completion of the national target so that resources such as labor, water and electricity were preferentially supplied to the iron and steel sectors. The consequences were that agriculture, light industry, and tertiary industries were seriously underinvested; social public affairs and urban development were subjected to varying degrees of negative influences due to a lack of concerns;

the pace of urban construction lagged far behind those of industrialization and urban population growth; and ultimately, economic and social development in the city was seriously hampered, and urbanization process stagnated.

The unreasonable iron and steel industry-oriented development strategy was soon corrected in the latter half of the "Second-five Year" period and the "Third Front Construction" period (1965-1978). In order to repair the imbalanced industrial structure, municipal government increased investments in raw materials and in the light chemical and machinery-manufacturing industries. A number of municipal enterprises (e.g. Lanzhou Power Machinery Factory, Gansu Agricultural Machinery Factory, Gansu General Machinery Factory, Lanzhou Instrumentation and Components factory, etc.) were set up at the beginning of the 1960s. Based on the considerations of balanced regional development and military security during the Cold War, China established a campaign between 1965 and 1978. A large number of machinery-manufacturing and military factories as well as research institutes located in the coastal cities were sponsored by central government to relocate (to "hide") in the western mountain areas. By virtue of its important strategic location connecting eastern and western regions, abundant mineral resources of its surrounding areas, and a solid and modern industrial base, Lanzhou was again selected as a key city and received massive national investment as well as the relocated factories and institutes from the east. At the time, Lanzhou was repositioned by central government as a national machinery-manufacturing base in the western areas.

By the end of 1978, Lanzhou had received a total of 35 industrial factories and 14 institutes and universities and introduced over

10,000 engineers as well as entourages from eastern coastal cities (Table 4.1). Under the planning of the Central Committee of "Third Front Construction", and urban layout characterized by obviously functional divisions first appeared. Referring to the initial edition of Lanzhou City Master Planning (1954-1972), most of the petrochemical enterprises as well as food and textile enterprises were settled in the Xigu district, electronic and mechanical enterprises were placed respectively in the Anning district and the Qilihe district, and metallurgical enterprises and a few light industrial enterprises were placed in the Chengguan district. This kind of urban layout, giving priority to industrial sectors and having significant functional partitioning, has a profound impact on urban development and urban fabric of the following decades (Figure 4.1)[2].

Table 4. 1 Relocated factories and aid projects from eastern regions in 1965-1970

Enterprise's name after relocation	Place of origin	Build date	Employ ee (Person)	Total investment (10,000 yuan)
Lanzhou Carbon Factory	Jilin City	1965	700	7500
Northwestern Paint Factory	Tianjing and Dalian Cities	1965	288	400
Northwestern Organic Chemical Factory	Tianjing and Qingdao	1965	71	250

2   According to the latest Urban Master Plan (2010-2020), Lanzhou urban space includes four functional areas, namely Heavy Industrial District in west area (Xigu District and Qilihe District), Cultural and Educational District in southwest area (Anning District), Central Residential District in middle area (Chengguan District) and Light Industrial District in east north area.

	Cities			
Lanzhou Miner's Lamp Factory	Fushun City	1970	124	505
Lanzhou Changtong Electric Wire Factory	Shanghai city	1964	260	290
Lanzhou Changhong Welding Factory	Harbin City	1965	111	89
Lanzhou Bearing Factory	Beijing City	1965	800	527
Lanzhou Changxin Electric Meter Factory	Harbin City	1965	214	111
Lanzhou Electrical Machinery Factory	Shanghai and Harbin City	1965, 1969	723	295, 1862
Lanzhou High Pressure Valve Factory	Shenyang City	1965	300	695

Source: Lanzhou Yearbook (1990) and Lanzhou Statistical Yearbook (1980).

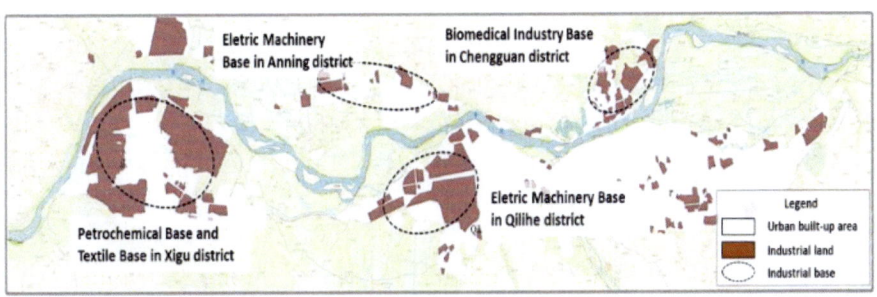

Source: drawn by the author based on the land use map that was published by Lanzhou Land &Resources Bureau in 2000.

Figure 4. 2 Lanzhou built-up areas and industry layout in 2000

Besides the petrochemical and manufacturing industries, the massive immigration of science and technology gave rise to rapid development in metallurgy, coal and electric power, and machinery as well as transportation equipment industries. In addition, other sectors, such as breweries, brush shops, food factories, and paint companies, along with industries associated with civic improvements flourished in the region. A modern industrial system (consisting of petrochemical, electric power, metallurgy, machinery, electronics and textile industries) began to take shape. By the end of 1978, Lanzhou's industrial output had grown to 3.63 billion Yuan, an increase of 240 percent compared within 1965. Additionally, the central government invested in and constructed a civilian airport, and several industry-specific railways, highways, bridges, as well as water and electricity projects in 1968 and 1969. Municipal engineering improvements, such as the construction of infrastructure, and the building of schools, hospitals and department stores were associated with the growth of the urban area. A new city relying on industrial bases rose up (Figure 4.2). This is particularly evident in the Xigu district. Accompanied by the immense industrial technology migrations, the urban population increased sharply from 1.34 million in 1963 to 2.06 million in 1978. The growth rate is 1.39 times higher than the national average.

4.1.2 Industrial decline and challenges
Population growth and urban space pressures
After the rapid industrialization and economic growth, the total urban population in the city rose to 3.62 million in 2012, up from 1.95 million in 1949, and this trend continues[3]. Based on the current annual rate of population growth (5.45%), this urban population

---

3   Data is collected from Lanzhou Statistics Bureau, 2013.

will reach around 5 million by 2020; by then, the urbanization rate will get up to 80%, which in accordance with official estimates require at least 420 km² of urban construction lands[4]. In fact, urban population density in Lanzhou since 2000 has exceeded 15 thousand people/ km², while the density in the core urban area (about 40 thousand people/ km²) is even higher than the ratio in Beijing (7,837 people/ km² in 2011) and Shanghai (9,589 people/ km² in 2011)[5].

However, being subject to the valley topography and scarce land resource, Lanzhou municipal government considers that the remaining undeveloped land in the city is far from able to meet the demands of such a fast-growing population. According to official statement, although the total area of the city is rich (13,300 km²), 85% of it is not suitable to be developed for construction uses (mountains, loess, hill and gully areas); only the 15% in the valley basin are exploitable[6]. Besides, the land-free allocation policy under the planned economic system results in an oversupply of industrial land in the urban area. Lanzhou municipal government has repeatedly accused of it in public. In accordance with the municipality's claim, a large area of industrial lands occupies the city center, and most of this has not been fully exploited; and most

---

4   Data is collected from Lanzhou City Master Planning, 2011-2020.

5   Data derived from Blue Book of Economic and Social Development in Lanzhou (2014-2015); Lanzhou Land Remediation Planning (2011-2015).

6   Data from: the website of China Investment Corporation & Urbanization and Industry Investment Co., Ltd., http://www.ztgfcz.com/showArticle.asp?id=124.

of industrial lands in the city center are idle and abandoned by the enterprises and institutions to which they were allocated far beyond their actual needs during the planned economy period. According to Table 4.2, there are 207.03 km² of land available for Lanzhou's future construction, while in 2009 the built-up area in the inner city has already reached 133.74 km². That means the remaining area is less than 75 km².

Table 4. 2 Potential construction land in Lanzhou built-up areas in 2009[7]

Total area of Lanzhou city (km²)	Area available for urban construction (km²)			Suitable amount of urban population (million person)	
	Total area	Current area	Remaining area	Suitable population	Current population
1055.12	207.03	133.74	73.19	2.02	2.33

Source: basic data was collected from Lanzhou City Master Plan (2011-2020) and calculated in accordance with the standard and methods[8] published by Lanzhou municipal government.

---

7   The state explicitly put forward "to stick to the deadline of cultivated land (1.8 billion acres) in order to ensure grain security" in the "Eleventh Five-Year Plan of National Economic and Social Development" in 2006. Therefore, since 2013, the unused land that can be developed each year in Lanzhou city is strictly limited below 1.5 million acres (10 km²), wherein the annual quota in urban areas is only 2,000 acres (1.33km²).

8   Total amount of available land for construction purposes is collected from Lanzhou City Master Plan (2011-2020). Population capacity is calculated according to the formula: $Cm=(100*QL*Im)/Rm$, in which QL is the total amount of construction land in the valley (km²), Im is the suitable proportion of residential land (%), and Rm is the land use standard of suitable land per capita (m²).

Source: photos were taken by author in 2013.

Figure 4. 3 Building density and traffic congestion in Lanzhou in 2013

According to this calculation result, the shortage of land resources is increasingly severe accompanying the rapid population growth. In fact, it has given rise to a series of problems, such as a serious housing shortage in the form of high prices and high-density, dilapidated living environments, inadequate infrastructure, and traffic congestion (Figure 4.3). Taking urban housing as an example, the average price per square meters in Lanzhou is higher than those over the same period in Xi'an and Chengdu—both of which are the capital cities in Western China and are adjacent to Lanzhou city[9]. Between 1996 and 2012, there was a tenfold increase of housing prices in Lanzhou, from less than 900 yuan/ m² up to 8,624 yuan/m², while the average monthly wage of employees was 3,300 yuan in 2012[6].

According to the standard of Lanzhou City Master Planning, Im (25%-40%) takes the value of 30%, and Rm (23m²-36m²) takes the value of 30m².

9   The GDP of both Xi'an and Chengdu is twice as many as that of Lanzhou.

Additionally, the traffic pressure in Lanzhou is particularly prominent. In the city, the road construction area per capita was 6.23 m² and accounted for 7.77% of the total construction land in 2012. The ratio is far less than the national standard in the provision of Urban Traffic Planning and Design Specifications (GB50220-95)[6].

Unreasonable layout and demand for new urban space
Due to several nationally invested projects and massive factories migrating from Eastern China between 1953 and 1971, a significant spatial change occurred in the city and led to its contemporary embryonic form. The whole city was divided into four parts, each of which was designed for the development of a professional industrial category. Urban space was primarily constituted by the four traditional industrial bases, namely the Petrochemical and Textile area in the Xigu district, the Bio-medical industry area in the Chengguan district, and the Electric Machinery area in both the Qilihe and the Anning districts. Surrounding these production spaces, plenty of workers' welfare housing and sporadic supporting facilities (such as affiliated hospitals, schools, canteens, commissaries, etc.) were jointly built by these factories as well as the municipal government.

This urban construction scheme giving priority to the production usages resulted in an urban layout characterized as a typical Soviet-Union-style. The city was characterized by a lack of a clear functional division in urban space; massive inefficient uses of industrial land occupying the inner city; sporadic commercial and residential lands interspersed among industrial spaces, and so on. This chaotic urban layout brought about serious urban challenges, such as deepening conflicts between local residents and industrial enterprises because of environmental degradation and industrial nuisance, and the insufficient residential and commercial space in support of fast population growth.

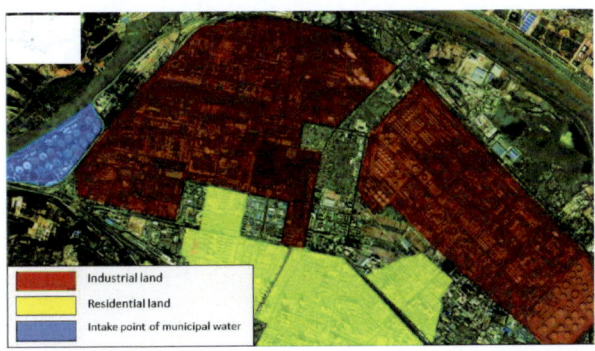

Source: picture was drawn by the author based on the remote sensing that was downloaded from Baidu Map.

Figure 4. 4 Land layout of Xigu district in Lanzhou around 2010

The affiliated spaces together with the production sites are collectively referred to as *"industrial yards"* (工业大院), or also known as *"danwei yards"* (单位大院), which were the basic elements in charge of the production and social management in Lanzhou. Figure 4.4 shows the land layout of the biggest industrial areas — the Xigu petrochemical industry base in Lanzhou around 2010. The industrial production area located in the upper reaches of the Yellow River is very close to the living areas and the intake point of the municipal water. The second picture shows the outlook of Lanzhou Petroleum Machinery Factory (Figure 4.5). This factory was constructed in 1958, and at that time, this area still was part of the urban fringe. Thereafter, the factory gradually built a great deal of workers' welfare housing and living facilities (e.g. schools, hospitals, cinemas, etc.) surrounding the plants and warehouses. With the dramatic increase in urban population and fast commercial development, this factory is now surrounded by a large number of new commercial and residential buildings. This factory is not a special case. Actually, similar cases are easily to be found in many places in the city. The rise of new urban spaces is constantly compressing industrial land, while plenty of industrial lands in the inner city in turn severely limit the expansion of new urban space. Competition for space between different economic sectors is becoming increasingly fierce. This situation is becoming more common, especially after entering the 1990s when the city was plagued by a number of economic problems, such as rising resource costs and industrial downturn, pressures from the service-oriented transition in national strategy, and so on.

Source: photo was taken by author in 2013.

Figure 4. 5 Mixed land-use of Lanzhou Petroleum Machinery Factory

Industrial pollutions and urban security

As a large number of heavy industrial projects (mainly petrochemical industries) settled in the upper reaches of the Yellow River between the 1950s and 1960s, Lanzhou has long been facing huge environmental risks. The city is sandwiched between mountains to its north and south that almost block the air flow during the entire year. Arid climatic condition and a lack of vegetation in the valley basin as well as on the northern and southern mountains easily cause a stable inversion layer in winter. The semi-enclosed valley terrain and arid climatic condition are not conductive to the outward diffusion of industrial pollution. In view of the fragile environmental capacity, in fact, Lanzhou is not an advantageous location for industrial placement, especially heavy manufacturing and chemical industries.

Source: Photos were downloaded from the website
http://news.focus.cn/lz/2013-04-23/3177422.html.

Figure 4. 6 Air pollution in winter of 2010 and chemical explosion of Lanzhou Petrochemical Company in June 2006

Since the 1970s, air, water and soil pollution incidents arose frequently. Between the mid-1970s and early 1980s, the Xigu district (i.e. the center of petrochemical industrial factories) suffered serious smog pollution every winter. However, all accidents had been concealed and were not revealed to the public until the late 1980s. Frequent pollution and safety accidents triggered a panic among residents, and the conflicts between local residents and industrial enterprises grew severe. Especially in recent years, the constant stream of pollution and explosion incidents made for negative media publicity for the city.

Confronted with the harsh criticism of the Ministry of Environment Protection of the State Council and the public pressures from the media and society, Gansu Province and Lanzhou's municipal government have to take action. They first put forward a "Blue Sky Project" since 2002. According to the official claims, initiatives of this multi-administrative project involve forcing seriously-polluting individual enterprises to close down or carry out technological transformation and equipment upgrades; limiting the number of new motor vehicles and travel frequency; shutting down partial coal plants and encouraging clean energy applications (e.g. conversion of public transportation to natural gas, and laying gas pipelines to residential areas), removing small scattered boilers and building central heating system; as well as planting trees.

Despite these measures, pollution is proceeding without an end in sight. One of the main reasons is that both provincial and municipal governments did not take practical actions in consideration of local economic growth and their own administrative performances. The latest contamination incident occurred in April 2014. The city's running water was polluted by benzene that leaked from an aging pipe at the Lanzhou Petrochemical Company. In fact, oil spills of this state-owned company have already caused two explosions in 1987 and 2002.

However, it is different from the previous social repercussions. The latest incident did not only cause a conflict between residents and enterprises, but it also intensified the contradiction between the municipal government and the polluted enterprise.

Source: Photos were downloaded from websites
http://news.xsjk.net/shrdxw/jdpl/20144/359043.html and
http://money.591hx.com/article/2014-04-11/0000299895s.shtml.

Figure 4. 7 Chemicals leaks and water pollution in 2014

In the official report[10] published in 2012, Lanzhou has been listed in the most polluted cities for many decades. Then in the new national environmental protection law published in the first half of 2014, environmental protection has become an important indicator to measure the performance of local officials. In face of these, Lanzhou's municipal government immediately denounced this company, and has shown unprecedented attention to the city's environmental pollution. In the 2015's municipal government announcement, the government argued that "the serious environmental pollution and poor living conditions accelerate the

---

[10] Towards an Environmentally Sustainable Future: National Environment Analysis of People's Republic of China.

outflow of talent and meanwhile undermines the attractiveness of the city for inward investment and migrations".

4.2 Lanzhou urban renewal strategy in new century

4.2.1 Calling on renewal: from industrial restructuring to inner city revitalization

The fast growth of heavy industries over half a century has brought the city a temporary prosperity and a rich heritage of industrial civilization. However, it has also caused various urban problems, such as the shortage of construction lands because of waste, irrational urban layout, environmental pollutions, and the consequent traffic congestions, soaring housing prices, and deterioration of the living environment. All these have to some extent hampered local economic growth and undermined urban competitiveness in terms of drawing inward investments, and moreover, they have indirectly brought negative impacts to the performance of local officials. Thus, it forces Lanzhou's municipal government to re-examine its past urban strategies and blueprints, urban planning, the means and approaches of governance, and so on, and to find out possible solutions.

In addition to the pressures from the city itself, the national economic transition also presented Lanzhou challenges to change. Since the late 2000s, the central government put forward a transformation plan in the entire country. This plan aims to cope with the urgent issues, concerning the slow crawl of the national economy, an aging population, the shortage in the labor supply, the ongoing rise of average wages as well as production costs, and so on. All these are considered the main obstacle to national economic and social stability since entering 2000. In the national plan, the central government advocated that "China's future economic growth should transfer its engine into replying on domestic consumption, population urbanization, and a service- and low-carbon-oriented industrial transition". That is to say, the country will shift its role from a traditional production state to a service-based and consumption-based state. The central government launched a series of mega urban renewal and infrastructure investment programs, involving the projects of shantytown transformation, highway construction, reconstruction of traditional industrial bases, and the revival of the New Silk Road. These projects, according to the official argument, are a preliminary scheme of the national plan and conducive to stimulating economic growth, increasing employments, improving the living environment, and promoting social harmony and national prosperity.

Although the state did not enforce localities to comply with these strategies according to its official documents, Lanzhou's municipal government still gave a timely response. To re-evaluate its changing political and economic situation makes the city quickly reposition its targets for local economic development and urban construction, to be in compliance with the blueprints of the central government. Their purposes are rather clear. One is that the municipality hoped to gain a competitive advantage, compared with that of other cities which were confronted with similar pressures to transform, in such a turbulent era that is full of opportunities. More importantly, it hoped its positive and timely response would enable it to be much easier to acquire the preferential policies and investments from the central government.

Given the political dependence on its higher authorities and the fast changing national economic environment, the central government's planning and strategies become the basis for the ideas of urban authorities to develop their strategies and plans. The first step that Lanzhou's municipal government adopted was to discard its traditional role as one of the most famous national petrochemical base since 1960s. It hoped to seek new economic opportunities those can help it throw off the outdated looks—the motors' roaring and the stacks of chimneys that were everywhere in old days—and bring it a refreshed and vibrant image. Rebranding city is an important tactic taken by the Lanzhou municipal government. The aim of this approach is to arouse a consensus in the society and establish a rational basis for its next actions. To this end, the Lanzhou municipal government repositioned the city brand twice. One of its repositions was to build an important "Logistics Center" in western China and a "Science and Education City" in 2000s. Another one was to create the city as an "Innovation Zone of Chinese Civilization" in 2013. Both of them are closely in line with the central government's strategies launched respectively in the late 2000s and 2010. Various slogans appearing in these years reflect a truth that is the strong willingness of local authorities to build a service industry-led and consumer-oriented city, by encouraging the development of commerce, real-estate, education, logistics and tourism sectors.

In order to achieve the targets of economic restructuring, Lanzhou was confronted with another issue, namely to provide sufficient lands for the newly-emerging economies in the city. The second step that the municipality of Lanzhou adopted is to rebuild the urban space. In consideration of the local special terrain as well as the national protection policy on agricultural land [11] , the redevelopment of existing construction land—i.e. rebuilding the lands for more efficient and intensive uses—was unanimously adopted by provincial and municipal administrative departments as a solution. The way of rationalizing the solution is to develop urban planning. In China, the municipal government is appointed by the central state as the main actor responsible for the development and review of urban planning; meanwhile urban planning is given an unchallengeable status at the legislative level. As such, it is the best way of materializing the government's conception, through which government's words and ideologies are working realities and are institutionalized as an accepted commonsense. After reviewing its previous City Master Planning and the introduction of new versions [12] since 2000, the government's strategies quickly attracted a wide attention of scholars and local media, and subsequently turned into a social consensus without too much controversy. The headlines, such as "Building seven roads in core area to ease traffic pressures", "Reconstructing Huanghe Building to create a new landmark in Lanzhou", "Promoting industrial enterprises' relocation and establishing the Fifth National Industrial Zone", "Rebuilding North Longkou as an international trade & logistics center", and so on have appeared frequently in local major newspapers.

---

11   The central state issued a stringent provision on agricultural land protection in 2007—i.e. "to protect the country's 18 million hectares of agricultural land against erosion".

12   In the third edition of Lanzhou City Master Planning (2001-2010) and the fourth edition of Lanzhou City Master Planning (2011-2020), urban renewal and new town construction are two most important targets.

4.2.2 Selecting renewal objectives: shanty town, urban village and traditional industrial area

After rendering this municipal movement and making it legitimized, Lanzhou city launched a gradual, government-driven, multi-faceted urban revitalization plan as the supplementary to regulate the behavior of the participants. First, it promulgated Lanzhou Land Use Planning (2006-2020) in 2005 and Lanzhou Land Remediation Planning (2011-2015) in 2010. These two plans were intended to clarify the main area for future urban construction and additionally aimed to formulate strict limitations on the land development processes, regarding the development approaches, the total amount of land development, and the end-use types of construction land.

Given the local reality that there is no sufficient non-construction land available for further urban construction, the Lanzhou municipal government has taken an alternative approach—i.e. shifting its focus from agricultural lands to the idle and inefficient-use lands of inner city. Another reason of doing so is in consideration of the increasing risks and costs. The central state has imposed a stringent law on farmland protection and ordered local governments to raise the compensation standards on the encroachment of agricultural land since 2005. As a result, the urban villages, shanty towns, old mining and industrial area (e.g. lands for industrial storage and production, and residential lands of the state-owned work units) became the key targets of municipal remediation and reconstruction (Figure 4.8). It is known as the *"Three Old Transformation"* (三旧改造) in some Chinese cities, such as Guangzhou, Shenzhen and Shanghai. In this way, the municipal governments solve the problem of shortage of land and cater its strategy to the central government's blueprint; and furthermore, they have successfully overcome the institutional constraints.

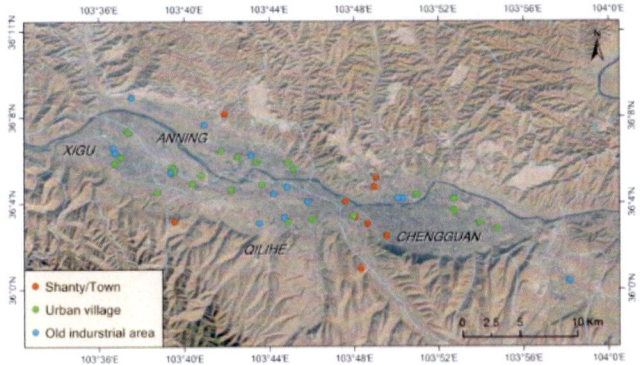

Source: drawn by the author on the basis of "Lanzhou land remediation planning (2011-2015)".

Figure 4. 8 Key areas of Lanzhou urban renewal

Similarly, how to make the new municipal strategy rationalized and accepted without controversy is the issue that Lanzhou municipal government has to face. The most effective method is still the formulation of urban planning. By highlighting the unreasonable land-use situation, explosive population growth, and the shortage of construction land as the biggest predicament that the whole society should urgently face in its government announcement, the municipal government immediately developed the solutions. In its detailed programs, land conversion from agricultural to construction lands (i.e. newly added construction land) is strictly controlled. In the Lanzhou Land-use Planning (2006-2020), the municipal government stipulated that the total amount of the newly added construction lands converted in rural areas should be limited within 12,513 hectares between 2006 and 2020, in which the exploitable area is limited to less than 3,500 hectares before 2010. Then it argued that urban land development should be conducted in a more economical and intensive way for a sustainable purpose, that is, by the means of increasing building density and raising the floor area ratio (not less than 4). Then, it argued that only urban renewal and reconstructing the inefficient-use lands for some highly value-added business and residential uses is the best and most effective approach to overcome urban disease.

Table 4. 3 Urban renewal policies and relevant planning of Lanzhou (since 2009)

Urban renewal policies	Year	Publishers
Notice on the Promotion of Real Estate Market	2010	State Council
Guiding Opinions on Promoting the Process of Shantytowns in Industrial and Mining Areas	2010	State Council
Opinions on Accelerating the Process of Shantytowns	2013	State Council
Reconstruction Plan of National Old Industrial Bases	2013	State Council
Guiding Opinions on Promoting the Intensive Use of Land Conservation	2015	National Land &Natural Resources
Implementation Suggestions on Shantytowns in Urban and State-owned Industrial and Mining Areas	2010	Provincial Government
Suggestions on the Central and Provincial-owned Industrial Enterprises Moving into the Lanzhou National Industrial Zone	2011	Provincial Government
Opinions on the Central Policies about the Construction of Lanzhou National Industrial Zone	2012	Provincial Government
Suggestions on Accelerating the Process of Shantytowns	2015	Provincial

and the Construction of Supporting Infrastructure		Government
Management Measures on Reform Planning of Lanzhou Urban Villages (trial)	2009	City Planning Bureau
The Fourth Edition of Lanzhou City Master Plan (2011-2020)	2010	Municipal Government
Lanzhou Land Remediation Planning (2011-2015)	2010	Land Resources Bureau
Suggestions on Further Accelerating the Process of Shantytowns	2010	Municipal Government
Opinions on the Support of Central and Provincial Governments on Industrial Enterprises Moving into Lanzhou National Industrial Zone	2012	Municipal Government
Measures of Industrial Enterprises Relocation and Land Reconstruction in Lanzhou	2012	Municipal Government
Opinions on Further Regulating the Implementation of Primary Land Development (trial)	2015	Municipal Government
Relocation and Reconstruction Program of Qilihe Old Industrial District	2013	District Government

Source: sorted by author based on the relevant documents published in official websites.

After defining the targets and a methods of future's urban development, the municipal government, joint by provincial government as well as relevant administrative departments at the local level, has immediately issued several particular embodiments since 2009 (Table 4.3). In these policies, it specified the specific initiatives and detailed standards regarding implementation (e.g. land planning, allocation of responsibility, construction measures, relocation compensation, project financing, financial support, tax incentives, profit distribution, etc.) so as to meet its targets.

171

After introducing a wide variety of urban construction planning as well as supporting policies, Lanzhou municipal government has finally enabled to implement its carefully-planned economic restructuring and redevelopment projects. These programs have provided strict limits to the subjects of responsibility, their obligation and tasks, the way of the participation, construction standards, and so on.

4.3 Policy innovation on land development: opening market and building a multi-cooperation framework

4.3.1 A strategic concept from central and provincial levels

Although Lanzhou municipal government as the vast majority of local governments have been empowered greater autonomy than ever before, administrative interventions from center government never disappear. Authoritarianism has long acted as the traditional core concept of Chinese governance, and is reflected in its strict hierarchical administrative organization and social structure. It derived from Chinese mainland-style civilization and patriarchal culture, has been considered as the most efficient way to mobilize people, organize production, and especially promote urban construction in its history. Even if the country has undergone reform and opening up, its hierarchically socio-political organizations still persist. Hence, the existing institutional arrangements and organizational structure will inevitably lead to a fact that local authorities continue to be subjected to the influence from central government.

This impact becomes nothing but increasingly implicit, shifting from a direct to an indirect way. On the one hand, the central government has abandoned its usual harsh tone of commands, while using more moderate discourses and policies to inspire local governments. On the other hand, the notion of "bottom-up" obedience almost pervades in all aspects of Chinese culture, latently or explicitly, overtly or covertly. The entrenched awareness leads to the local authorities' habitual compliance with superiors, as well as their reliance and obsession on the central commands. The implicit and indirect influence can be captured in the process of municipal governments choosing strategies for local economic and urban development, as mentioned in last section. It can be also noted in the specific actions of local governance, regarding the policy-making in urban renewal. This last makes municipal governments follow the rules of the state-dominated framework regarding an entrepreneurial urban development step by step.

A national strategy from the central government

Given the country's failed experience of urban renewal before the 1990s, a market-oriented operation, reducing the intervention of local governments, and redefining local authorities' duties were repeatedly mentioned at the national level. For instance, the State Ministry of Housing & Construction at first put forward in 1993 that:

> Urban renewal in the future 'should' be carried out with the assistance of real estate market; meanwhile, the tasks of local governments are to improve the abilities of municipal financing and to create opportunities for the involvement of private capitals in urban construction—extracted from the

working report of the Future's Construction and Maintenance of Urban Housing[13].

Then, in the latest national renewal plans—i.e. Guidance on the Promotion of National Shantytowns, and the Transformation Planning of National Old Industrial Bases—formulated respectively in 2012 and 2013, the central state gave further suggestions on the duties of local governments. According to the central state, local governments are asked to streamline their organizations and simplify administrative examinations and approvals, including the procedures and the contents that involve. Besides, local governments are expected to open more investment areas to the public, and simultaneously to lower the investment thresholds. Moreover, the central government suggests local authorities to provide an enabling policy environment for financing innovation, so as to attract a wider participation of social capital as far as possible. For instance, the State Council has pointed out in its official document issued in 2016 that:

If conditions permit, cities/regions should be committed to innovating new financing methods of urban construction...— extracted from the Opinions of the State Council on the Further Promotion of New-Type Urbanization[14].

---

13   The meeting was held by the State Ministry of Housing & Construction in Xian in 1993.

14   Source from: http://www.gov.cn/zhengce/content/2016-02/06/content_5039947.htm

This national document encourages that the renewal projects should preferably be accomplished with the support of multi-channel funding sources. Except for a small amount of special funds from the investment of the central government and the financial allocations of localities, most of the funds are required to be raised through a policy bank (China Development Bank) and commercial banks.

In addition to providing advice on financing, the State Ministry of Housing & Construction formulated a monetary system on relocation compensation in November 2001. In this policy, the ministry pointed out the method of compensation in kind for relocated households (the same size housing, double or triple areas of housing) was no longer advocated, being replaced instead by monetary compensation. According to the official statement, the establishment of this monetary compensation will alleviate the financial pressure for central and local governments, by means of capital markets. Although not explicitly stated, another purpose of this initiative is clearly to stimulate the sluggish real estate market.

However, in the subsequent ten years, the application and promotion of monetary compensation in the country was rather slow. It is in large part because of a lack of legal policy to clarify the sources of compensation funds and the manner of compensation, as well as the responsible subjects who implement the compensation. To this end, the State Ministry of Housing & Construction had to add in 2012 that local governments "must" play a major role in coordinating relocation and compensation, including increasing the proportion of monetized resettlement and formulating compensation standards, and if necessary, to assume the obligation of compensation, such as the monetary subsidies for poor urban households [15] . Additionally, the administrative compensation funds from both national and local budgets were totally banned in law.

Paraphrases from the provincial government

Provincial governments are the first subordinate executors of central commands and grantors of national projects funding, whose basic tasks are to coordinate the work of the counties and cities within the regions, to execute guidance and supervision on local implementation of national projects, and more importantly to allocate the special funding to cities and counties—within their administrative scopes—in accordance with the central government's plans. With respect to urban construction, the central government proposes national strategic visions, and entrusts the provincial government to convey them to the local level. The administrative power of provincial governments in intervening urban affairs has been greatly weakened. Only some supervisory power is left over. Thus, for the local level, decentralization from the provincial level is far beyond that of central government.

After the promulgation of national programs on urban renewal, the Gansu provincial government—the superior administration of Lanzhou municipal government—has launched several policies since 2010. In the policies (Table 4.3), it promised policy support and the necessary financial assistance and offered advices to municipal governments. These can be summarized as follows:

---

15   Source from the official website of the State Ministry of Housing & Construction: http://www.mohurd.gov.cn/zxydt/201509/t20150902_224630.html.

(1) The Gansu provincial government promised to invest a certain amount of initial funds in Lanzhou urban renewal projects (e.g. the provincial government invested 500 million Yuan for the infrastructure establishment of the Fifth National Industrial District to support industrial enterprises relocation), and meanwhile called on a broader involvement of multi-stakeholders (e.g. administrations, commercial banks, trust companies, real-estate developers, relocated enterprises and residents); (2) the provincial authority recommended the application of public-private partnerships ("PPP") and encouraged private bodies to play a leading role in the renewal process, especially reputable real estate developers; (3) it urged Lanzhou municipal government to devote itself to improving the local credit system and to ensure equity financing and debt financing.

4.3.2 The real demands of localities

As described above, municipal governments are urged/suggested by national legislation (in the Eighteenth National Congress) and provincial recommendations to provide various services to facilitate marketized operation in urban (re)development and to regulate the market through legislative authority. However, the market-oriented operation of urban development cannot be only attributed to the needs of the central authorities, but seems to be favored by municipalities according to local demands. Influenced by the discourses about "administrative decentralization", the concept of a "small state, and big society", which means that government being gradually withdrawn from direct participation in the social production and reproduction processes (He and Wu, 2009, p4), are gradually accepted and approved by local political elites in all Chinese cities, and Lanzhou is no exception. Meanwhile, given the failure experiences of the 1980s and 1990s—during which governments and state-owned enterprises swept all urban construction tasks, leaving the central state and localities to bear onerous administrative and financial burdens, local authorities eagerly desire to get rid of the role of "omnipotent agents" (He and Wu, 2009; Wu and Yeh, 2007) instead acting only as a coordinator, or more precisely as a manager in the urban (re)development process. Thus, making full use of market regulation and encouraging a wide-ranging involvement of social communities by reducing government functions as well as burdens are also motivated by the current needs of local authorities.

4.4 Entrepreneurial attempts of municipal government

4.4.1 Outsourcing administrative functions

Delegation of authority: establishing municipal "subsidiaries"

In order to simplify the administrative examination and approval the procedures of redevelopment process, Lanzhou municipal government has called for a self-restructuring among its municipal administrative units. On the one hand, it insisted that Lanzhou municipal government and its subordinate administrative units must be the primary designers and regulators of the renewal movement, relying on their monopolies on urban land property and the rights of formulating development strategy and urban planning. They are asked to engage in the formulation of the overall renewal plans, conduct a supervisory role in the renewal process, and be in charge of the assessment of project results. Generally, the responsibilities include formulating urban and land-use planning, deciding the participants as well as their specific functions and developing the fund-raising schemes, monetary compensation methods and construction standards of new projects as well as policies concessions, and so on.

On the other, the municipal government urged itself as well as the subsidiaries to transfer partial duties to its lower-level administrative departments. First, the municipal government delegated part of the autonomy to district government since 2005. District governments are given the rights to design specific projects and introduce investors. For example, the redevelopment project of the traditional industrial base in the Qilihe district was designed entirely by the Qilihe district government in 2013. Although the relocation compensation measures, and the issues related to where and when enterprises will move have been already proposed by the municipal government, the development projects and final land-use types are in fact largely dependent on the negotiations between district governments and work units/ enterprises (for those who intend to redevelop the land by themselves after the relocation). To some extent, the district governments in Lanzhou have switched from a passive executor to the implementers who hold partial decision-making authority.

Then in 2014, Lanzhou municipal government decided to further delegate the decision-making power related to urban construction and land development to district governments. The municipality at first selected the Xigu district as a pilot and granted it the tasks related to infrastructure construction and the management rights within the scope of its administrative region. Meanwhile, the Xigu district was asked to be responsible for the fundraising, organization and implementation of all renewal projects, as well as the project approval and acceptance, which were assumed by the municipality in the past. In addition, municipal government awarded tax incentives to the district government. The municipality promised to refund all city maintenance taxes that were submitted by the Xigu district in the last year to the district government, and meanwhile agreed to return all land revenue to the district finance bureau.

This shift as the municipal government wished has helped itself withdraw from the onerous burden of urban construction and management as experienced in the past. By sacrificing partial municipal finance as the expense, the municipality now only needs to be in charge of drawing up planning and politics, and to perform a delicate balance among district governments. District governments are given a certain degree of authorities, and this change has mobilized their enthusiasm for engaging in local construction. They do not only perform as a direct operator, but more like a subsidiary corporation of the municipality, who has independent liability but is required to be responsible to its parent company.

Transverse decentralization and the new administrative agents

In addition to the "longitudinal" decentralization of administrative functions, Lanzhou municipal government has spun off the tasks related to urban construction from its administrative functions and set up several specialized agencies since 2000. These agencies exercise the duties regarding land reserves, asset management and project construction, and project financing on behalf of the relevant government departments. Acting as the municipal agencies, Lanzhou municipal government offers them with funds, asset-backing, and policy supports, while allowing the agencies to undertake the development and operation of state-owned assets and income generation. In addition, the municipality promises to grant a temporary debt-funding when the agencies are confronted with a shortage of funds, and/or to bear partial debt obligations when they are caught in a debt-servicing difficulty. Since the establishment and post-operation of these agencies are largely dependent on the financial allocations or the injected municipal land, stock equity and other assets of local governments, they are actually a kind of alternative state-owned enterprises. That is to say, they often exist in the form of the special independent entities with an administrative nature. The inescapable duties that the agencies undertake are to accept all the construction tasks requested by the municipality (including fund-raising, construction, project operation, etc.), and to prorate their earnings with the municipal government after the project development and operation. Owing to the state-owned property, most of them normally engage in the investment, financing, construction and operation of quasi-operational and public projects which are sponsored by the municipal government, but a few concurrently get involved in the operating business.

For instance, agencies set up through the capital injection in the form of land, real estate and other municipal assets (water, electricity, and so on), are managed and supervised by the Municipal Finance Bureau and Municipal Land Resource Bureau. Such agencies (e.g. Lanzhou Land Reserve & Investment Centre "LLRIC") have obvious administrative color, since the sources of assets and their financial operation and management have to get the approval of Lanzhou municipal government as well as other relevant administrative departments. All of the gains they created are required to be turned over to the Municipal Financial Bureau, in addition to a small amount of necessary expenditures.

As for other types of agencies (or the municipal companies) whose establishments are attributed to the government-funding, but run under the supervision of State-Owned Assets Supervision and Administration Company (SASAC), the case is slightly different. Due to a reliable financial base and the granted autonomies, their business operations and profits generation are normally dependent on the market rules as well as their own abilities. They have every right to retain the profits at the local level for further investment and business operations, except for turning over a small portion to the municipal financial bureau as the sales tax. These agencies include Lanzhou Industrial Development & Construction Company (LIDCC), SASAC Lanzhou Construction & Development Company (SLCDC), and Lanzhou Urban Development & Investment Corporation (LUDIC).

4.4.2 Market-oriented experiment

In addition to outsourcing part of the administrative functions by establishing semi-administrative agencies, the market-oriented experiment at the local level is another important measure that the municipal government has developed in response to the national politico-institutional changes. In the renewal planning and relative policies of Lanzhou, the market-orientation is embodied in three aspects, namely adopting a property-led development, establishing an open tender platform, and encouraging the innovation of financing.

A property-led development

The municipal renewal policies in Lanzhou indicate that a reliance on the property-led development with involvement of private sectors functions as a basic driving force in promoting the city's redevelopment process. Since 2010, Lanzhou municipal government has issued three initiatives in order to achieve this target. In the municipal document, entitled "Opinions on the Support of Central and Provincial Governments on Industrial Enterprises Moving into Lanzhou National Industrial Zone (2012)", for instance, Lanzhou municipal government stipulated that reconstruction of most industrial lands in the future should be redeveloped for commercial and residential purposes; and additionally, that the relocated enterprises are permitted to conduct the reconstruction on their own (with assistance of their real-estate subsidiaries or other private developers), or transfer the land to real estate developers directly in the market.

The second significant policy is the "Municipal Suggestions on Further Accelerating the Process of Shantytowns" issued in 2010. In this document, it is stated that, "the administrative departments in the city should devote themselves in the policy design, so as to stimulate the enthusiasm of real estate enterprises' participating in urban renewal, especially to attract the large powerful and reputable property developers". Furthermore, in order to ensure the total amount of new commercial housing, the Lanzhou municipal government set a limit on the proportion of affordable housing and low-rent housing. In the shantytowns projects, for example, the proportion is asked to be around 30% of the total construction area, while the rest 70% is available for the business development purpose. About the redevelopment of these projects, SLCDC is commissioned to raise funds and to build affordable housing and low-rent housing, and the state-owned enterprises and institutions are primarily responsible for the financing housing. The rest commodity housing (around 70%) is fully contracted to private property companies. It is particularly noteworthy that, in the latest document—"Suggestions on Accelerating the Process of Shantytowns and the Construction of Supporting Infrastructure (2015)" issued by the provincial government in 2015, the ratio of financial housing is lowered again to 10%. That means the construction of commodity housing will account for around 90% of the city's shantytowns projects. Compared to the shantytowns, the ratios of non-commodity housing in the transformation projects of urban villages and old industrial areas are even lower.

Shantytown transformation and "life in peace"

Source: Lanzhou daily, http://rb.lzbs.com.cn/html/2015-11/27/content_1431406.htm; pictures are downloaded from http://www.daneiedu.com/FCZC/523208.html.

Figure 4. 9 Municipal propaganda posters of Shantytown transformation

The property-led mode as the main approach of urban redevelopment is chosen by the municipal government for the following considerations. First, it hopes to utilize urban land as an initial capital and then to mobilize private capitals to stimulate land appreciation, through transferring and upgrading the old, deteriorated inner-city neighborhoods into high-profit communities, such as luxury apartments, office buildings, and commercial facilities. Second, as mentioned earlier, this initiative is conductive to promoting the city's image and improving people's lives through infrastructure upgrades, and thereby attracting inward talent and investment and increasing local revenue, as Lanzhou municipal government expected (Figure 4.9). Third, given the failure of previous experiences between the 1970s and 1990s, the municipality prefers to seek social capitals as a partner to jointly share burdens and also the risks, rather assume the entire responsibility alone.

Even though, there have been plenty of state-owned developers in the city, which emerged with the formation of work-units system in the planned economic era[16]. Private developers are chosen by Lanzhou municipal government as the ideal candidates for several considerations. The most essential goal of private property developers is to pursue a profit, especially with a view to short-term returns, by using finance, land, building materials, and labor that they have access to. Their usual practices are to invest in land, and then to produce or improve buildings and facilities on the ground for a return of profit. Maximizing profit through the land and buildings' appreciation happens to coincide with the targets of

---

16   In the era of planned economy, most work units had their own construction departments. Many of them are still in operation. In addition, Lanzhou municipal government has established several state-owned companies to be in charge of urban construction.

the municipal government. In addition, these professional development companies are normally in a strong financial situation, meanwhile possess extensive experiences of engagement in capital operation and project marketing compared with the state-owned developers.

Open tenders through the public transaction platforms

In the municipal renewal policies, the second experiment about market-oriented shift is to establish the open and competitive platforms for project tenders. There are two reasons for the municipality conducting this shift. One is to curb the collusions of officials with developers. The municipal government wishes to expose the transactions' procedure on a relatively transparent platform, so as to avoid the non-compliant black-box operations— i.e. administrative rent-seeking corruption and irregular transfer behaviors of officials at lower administrative levels, which were rather common in the past. The second purpose is tantamount to share the projects' information to the eligible economic entities at a wider scope. This practice may help the government get a higher tender price of projects (or land) by recruiting more bidders.

(1) A land tender platform

To this end, Lanzhou municipal government built up a public platform, named Lanzhou Land Reserve & Investment Centre (LLRIC) in 1999. It is used as a non-profit agent being only loyal to the municipality as well as Lanzhou Land Resources Bureau (LLRB). In addition to acting as a platform for land transactions that makes these procedures more transparent, LLRIC simultaneously assumes the responsibility of regulation in land market for Lanzhou municipal government. LLRIC deputizes for the municipal government to control the total amount of land supply and adjust the urban land price. The purposes of this control are, on the one hand, to avoid an excessive supply of land, which often results in the inefficient use and land wastage; on the other, to ensure the municipality could always access to higher land revenue by tightening the land supply in the city.

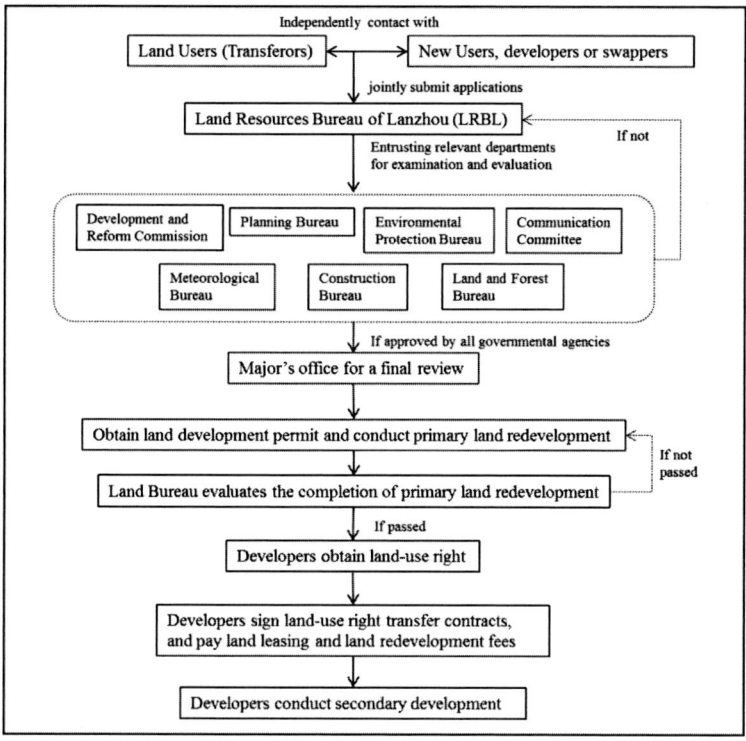

Source: drawn by author.

Figure 4. 10 Land transfer procedures in Lanzhou before 2000

Even though LLRIC has already been established in 1999, land transaction in the form of bidding and auction had not yet become a mainstream in a long period. In fact, most common practices during this period were analogous to that between 1970s and 1980s (Figure 4.10). Urban land users—mainly enterprises and institutions, especially the state-owned entities—were allowed to look for new land applicants by their own or under the assistance of LLRB, and to exchange their land with or donate to the applicants (with the permit of the municipal government)[17].

Since 2005, however, the Lanzhou municipal government introduced a new policy, in which most urban lands that are planned to be redeveloped are required to be handed over to LLRIC (Figure 4.11). In the policy, it is simultaneously regulated that all land transactions should be only carried out through open tenders. Besides, this policy made a clear distinction and definition of the roles and responsibilities of different governments and the specific administrative departments. Meanwhile, it set out the land transfer procedure in detail. In the policy, the Lanzhou municipal government and governments at lower levels are in charge of proposing objectives and plans of urban renewal. Afterwards, governments ensure Lanzhou Urban Planning Bureau to develop a land-use planning as well as relevant planning. The municipal government empowers the state-owned investment and development companies to carry on land acquisition. Then, it entrust the LLRIC for information disclosure and land transaction. After paying off land transfer fees as well as taxes, developers obtain the land and are allowed to conduct redevelopment by

---

17   The activities were supervised by LLRB before 1999 and by LLRIC after 1999.

themselves, or transfer the land use rights and/or the reconstruction tasks to other applicants again through LLRIC.

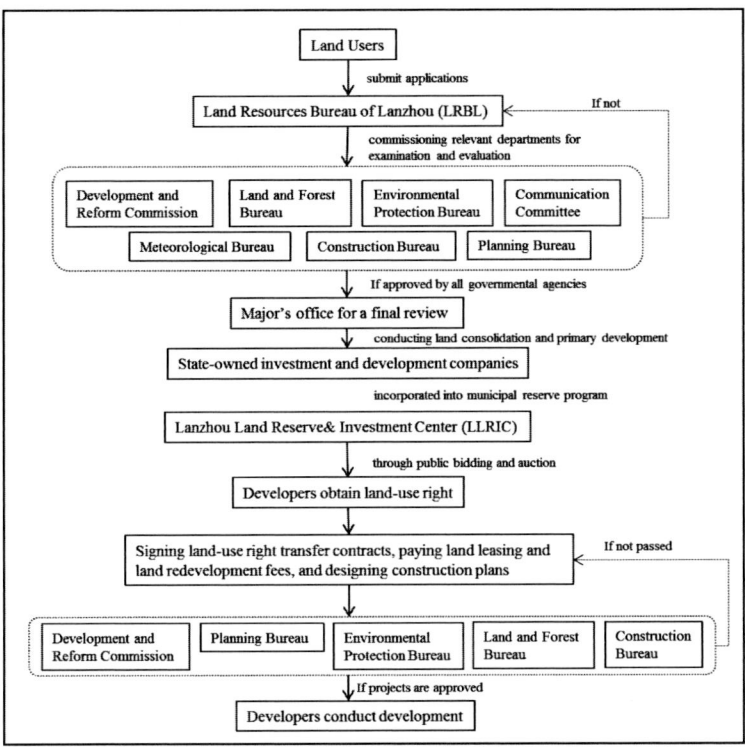

Source: drawn by author.

Figure 4. 11 Land transaction procedures in Lanzhou since 2005

Between 2009 and 2012, there were 41 industrial enterprises completing their relocation. All the land (about 2.67km²) was submitted to Lanzhou Urban Development & Investment Corporation (LUDIC) first for remediation and primary development, and these then were turn over to LLRIC. By the end of 2012, about 0.08km² lands have already been auctioned and traded by real estate developers for business and residential uses[18].

(2) A project tender platform

Lanzhou Investment & Trade Fair (LITF, "兰洽会")[19] is another platform that the municipal government openly call for participants (investors and developers) of urban (re)construction projects. The procedure of project bidding is similar to the process of land transactions mentioned above. First, the Lanzhou municipal government appoints LUDIC to operate in the land consolidation and the primary construction of infrastructure. Afterwards, the municipal government or sometimes the municipal construction departments are in charge of designing the specific redevelopment projects and then entrust district governments to make these items public on annual LITF. Finally, the district governments determine the project investors and developers by way of open competitive bidding and sometimes via negotiations.

18   Data derived from the website: http://lzzxqy.com/news/aspect/62.html; Lanzhou Urban Development & Investment Corporation (LUDIC) is set up by Lanzhou Municipal Construction Committee in 2000, and is mainly responsible for municipal project construction, asset management, and financing development, etc. It is one of the state-owned investment and development companies.

19   In order to promote local investment, Lanzhou municipal government set up the Lanzhou Investment & Trade Fair in 1993, which assumes the responsibility of the promotion of local culture, attracting inwards' investment, etc. LITF now is one of the biggest investment platform and an important trade fair in Western China. Since 1993, the total annual investments that Lanzhou municipal government introduced through the LITF have increased from 660 million yuan in 1993 to 612.9 billion yuan in 2014.

Different from the land bidding platform (i.e. LLRIC), LITF normally displays more renewal projects for which governments have already developed specific reconstruction plans. Rather than an open bidding platform, LIFT performs more like a kind of trade fair for the negotiations between business and governments. The items (or the areas to be developed) displayed on it involve those that are highly difficult to be reconstructed and/or show a lack of potential appreciation for the land and ground attachments. For instant, it displays some projects located in the urban fringe or the projects in the city center but with high costs of relocation compensation. This type of items is normally planned to be redeveloped for non-profit civil purposes, such as affordable and resettlement housing. In order to force the involvement of private developers, these non-profit public projects are usually bundled together with a few commercial projects. Obviously, it is unlikely to be consistent with the interests of private developers. As a result, although adhering to the public tender as a basic principle, district governments tend to consult with developers before or after the tender, and provide preferential policies (e.g. tax incentives, or the promise of some unplanned projects), with the tacit approval of Lanzhou municipal government.

On the 20th LITF in 2014, the Lanzhou municipal government and district governments jointly launched 57 projects involving industrial upgrades and restructuring, tourism development, inner city renewal, infrastructure construction, and so on. Thirty-five items of them are about urban renewal, and 20 of them relate to the shantytowns and construction of affordable housing and settlement housing. These renewal projects cover an area of about 0.6 square kilometers, accounting for 33% of 70 renewal projects (about 1.8 square kilometers) launched by local governments since 2012 [20], and the vast majority of these projects were reached through cooperation via consultations.

Innovation of multi-channel financing

The third market-oriented feature of municipal government's initiative is to establish multi-source financing. Unlike the previous practice in which the construction funds rely entirely on the appropriation of central and municipal governments, urban renewal since 2000 is operated in a capital-oriented way by means of a broad participation of social capital. The new financing innovations were designed by Lanzhou's municipal government in response to the imbalance between the rapid growth of local public investments and the shortage of funding. Between 2000 and 2004, the annual growth rate of fixed asset investment in Lanzhou was 10.5% while local revenue maintained only 7.2% growth[21]. Since 2005, this gap is widening. The growth rate of fixed asset investment is more than 20%, and local fiscal revenue is around 10%[22]. To solve this dilemma, Lanzhou's municipal government has set up several State-owned Financing Platforms (SFPs), whose main task is to assist the municipal government in raising funds for

20   Data are from the websites:
http://xbsb.gansudaily.com.cn/system/2014/07/06/015080614.shtml, and
http://gansu.gscn.com.cn/system/2014/01/04/010563340.shtml.

21   Data are from China Regional Economic Statistical Yearbook, 2001-2005, and Lanzhou Yearbook, 2000-2004

22   Data is from Blue Book of Lanzhou Economic and Social Development, 2008-2009

urban construction, such as LUDIC established in 2000[23], Lanzhou State-owned Assets Management Company in 2002, and Lanzhou National Capital Investment Construction Co., Ltd in 2007.

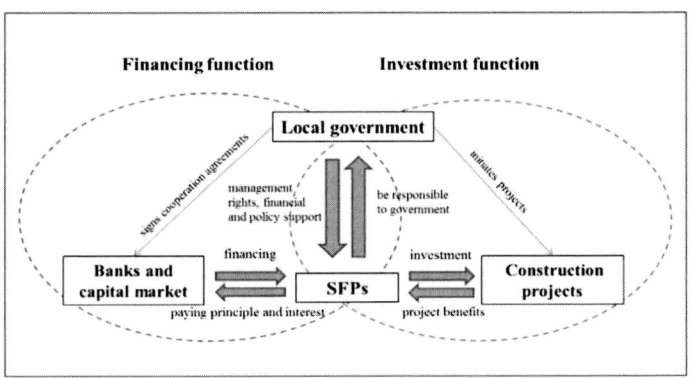

Source: drawn by the author.

Figure 4. 12 Relationship between SFPs and local government in funds transfer process

Specifically, the financing and operation mode of SFPs involves three cases (Figure 4.12). The first way of raising fund is that SFPs take loans from financial institutions (mainly National Development Bank and few large commercial banks) under the premise that the Lanzhou municipal government provides a credit guarantee for SFPs and signs a cooperation agreement with these institutions. After that, the loans are used as a special fund by SFPs for urban renewal projects designated for the purposes of relocation compensation and resettlement of enterprises (and

23   LUDIC and LLRIC were merged in 2005 and managed by Lanzhou Land and Resources Bureau.

individuals) and the costs of land leveling and primary infrastructure construction. The second funding approach is that SFPs jointly with few State-owned Trust and Investment Companies develop trust programs and issue trust products after the municipal government has signed a cooperation agreement with these companies. The operations after that are the same as the one in the first case. The third funding source of SFPs comes from the bank loans by mortgaging the land waiting to be redeveloped. In this case, Lanzhou's municipal government at first entrusts Lanzhou Urban Planning Bureau to determine the final land use and regulate the construction conditions. Then, LLRB evaluates the land according to the project planning and the potential for appreciation and defines the benchmarking transaction price. Afterwards, SFPs mortgage the land (or other lands with an equal value) to the financial institutions as security for loans.

These three fundraising approaches have, to some extent, helped the municipality solve the funding shortage of urban construction but have also brought the municipal government risks. SFPs act as an intermediary between governments and the capital market. Their financing processes are ostensibly regulated by the municipal government, in which all loans are secured by the municipal government by offering credit guarantees or by providing land assets as security. However, as the number of urban renewal projects grows and increasingly scarce land assets are unavailable for more mortgages, the capacity of municipal government to assume the debt is gradually reduced (Figure 4.13). Because SFPs' financing operation is closely related to the government's credit and the remaining urban land assets in the city, it thereby makes the financing more and more difficult. The SFPs that were initially set up for running non-profit public projects started to get involved in some profitable projects (such as real estate projects), since there is a lack of explicitly legal provisions of government for limiting such activities. Such acts of SFPs make the municipal government take on increasing potential risks. Most importantly, the funding source of this financing mode is single, in which financial institutions (mainly banks) are the main fund's providers, while other social capital in the form of securities, bonds, trusts, and other financing sources are not involved. Hence, it is obviously hard to provide stable and adequate funding for the government.

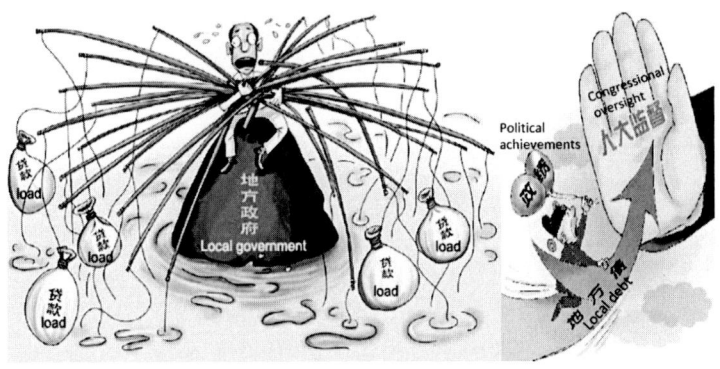

Source: China Economic Times,
http://lib.cet.com.cn/paper/szb_con/133127.html; Sina Finance.

Figure 4. 13 Local governments with dilemmas of credit guarantee

In order to broaden the sources of financing, a new way of fundraising has been designed after Lanzhou's municipal government issued a policy—"Interim Measures of SFPs in the Process of Industrial Enterprises Relocation" in 2008. In the policy, the issuance of corporate bonds is admitted by the municipal government as an experimental approach to raise finds. However, in the policy, SPFs is appointed as the solely responsible entities to operate this process; meanwhile, the bonds are limited to be only available to several large-scale companies and financial institutions. Its aim is to be sure that the fundraising procedure is always under the supervision of the municipal government.

Although having been aware of the irregularities and chaotic operations in the fundraising and project development processes, the municipal government did not make a clear limitation to it. Under the condition of financial constraints, the municipal government has no choice but to do so. It is because that the total amount of fundraising is significantly related to the progress of urban construction and the engineering of city image. It is thereby related to the administrative performance of local officials. However, the high risk of local financing is completely contrary to the original intention of the central government. In 2014, the State Council issued a document, entitled Opinions on Strengthening the Debt Management of Local Government, which aims to control the chaos and avoid the irregularities that happen in the financing process of local governments throughout the country. This central policy primarily involves the aspect of limiting administrative intervention. It calls on putting an end to the administrative financing guarantee to reduce the risk of government liability and on broadening the financing channels to attract more idle capital from society for all types of urban construction projects, including non-profit, semi-profitable and profitable (e.g. commercial real estate) projects. In particular, the roles and responsibilities of municipal governments are stated as follows.

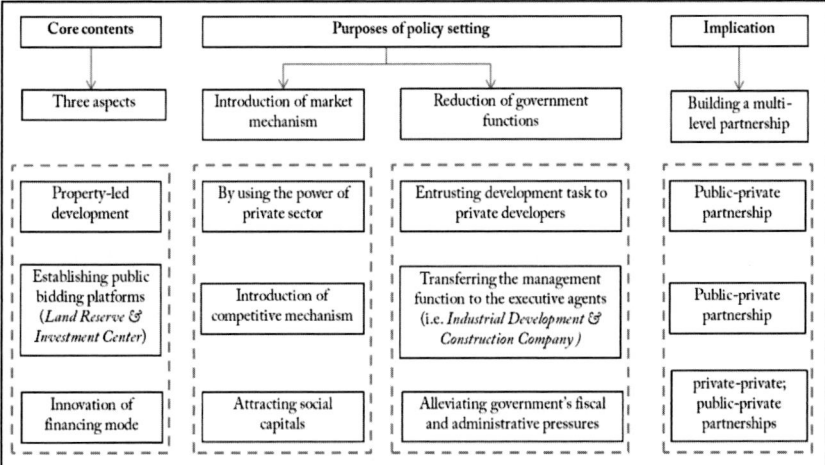

Source: drawn by the author.

Figure 4. 14 Local market-oriented experiments and its implication

First, municipal governments are urged to empower the management rights of SFPs to the SFPs themselves. That means the operating method of SFPs is switched to an independent operating mode. All SFPs are asked to be re-organized as independent companies in which the state-owned assets make up a large proportion. Municipal governments withdraw from the task of providing loan guarantees while SFPs are asked to seek financing with their own financial strength and credit.

Second, in the national policy, all financing projects are categorized into several types, and it is stipulated that each of them should obtain financing through different means. For instance, non-profit projects (e.g. affordable housing, roads, land reserves, etc.) are directly funded by municipal governments or through financing bonds issued by the municipality; semi-profit projects (e.g. water, gas, waste disposal, etc.) are to be operated under the joint cooperation of public departments, SPFs as well as private companies/developers (by establishing a kind of public-private partnership, i.e. "PPP"); the financing of profitable projects (e.g. commercial real estate) is to be no longer borne by SFPs as well as municipal governments, but rather is to be dependent on other channels, such as traditional ways of debt financing, and the innovative approach of private-equity holdings. The central intervention aims to regulate the behavior of local authorities through a detailed breakdown of project types and their financing approaches. However, it also leads to several contradictions. In accordance with the policy, the non-profit projects are no longer allowed to be operated by means of social financing and social capital; meanwhile local fiscal incomes are designated as the sole source of funding such projects. The limited local revenues are often not sufficient to support these projects.

4.5 Summary and discussion

This chapter examines local government transition, urban renewal, and governance practices through a case study of Lanzhou. It explores the contradictory characteristics existing in contemporary Chinese urban governance regimes. In Lanzhou, urban renewal strategies and municipal policies with regard to economic restructuring and urban construction are inevitably subjected to the central government's influence. Some of the impacts arise directly from regulations, policies, and planning blueprints issued by the central level. For instance, the central state has promulgated policies on the regulation of real estate, national standards of urban construction, a monetary compensation mechanism regarding the relocation of residents and enterprises, and several states' plans concerning shanty towns and the transformation of traditional industrial areas. Other effects are indirectly exerted by the center on localities, as through rendering problems and discursive mobilization. The central government advocated national targets for "shifting the state from a traditional production country to a service-led and consumption-driven country" and adopting a "small-state, big-society" approach. Its purposes are to raise consciousness regarding the urban living environment and consumption and to encourage the both local states and civil society to take responsibility for themselves

Due to these direct and indirect influences, local strategies maintain a high degree of consistency with the central government's goals. Even though, it should be recognized that some initiatives are made by localities for their own purposes. In pursuit of a win-win goal, the central and Lanzhou governments have formed a tacit strategic alliance around urban renewal. The goals are to cut administrative expenses on social welfare provision and infrastructure construction, promote urban housing commercialization, attract investment in real estate, and encourage a broader participation of social capital in urban construction.

In this context, the municipal authorities behave more like enterprises with a goal of maximizing their interests in the market, employing their means of administrative control and the monopoly on land, urban planning and policy-making to achieve that goal. Owing to the uneasy coexistence of decentralization and recentralization, however, the scope of localities' economic activities and their actual operating autonomy are open to question. Given that political careers remain tightly controlled by the central state, local politicians' strategic choices, decision-making, and the approaches by which they manage their cities are always subordinate to the core interests and strategic targets of the center.

Regarding urban construction, even though a broad consensus over outsourcing administrative functions of municipal governments and a market-oriented experiment has been reached between all levels of government, the real-world implementation of this reform has been far from smooth. This can be attributed, to a great extent, to the coexistence of market mechanisms and a planned economy system, and the resistance from that employing traditional managerialist governance to new entrepreneurial modes. The Lanzhou municipal government, on the one hand, tried to carry out the market-oriented reforms in compliance with the requirements from the central government; on the other, it took some sub-standard behaviors that are against market rules. For example, it conducted open bidding for public projects meanwhile adopting the private negotiations in some special cases; and the Lanzhou municipal government encouraged the commercial financing of SFPs, while simultaneously providing risky credit guarantees for the private land users. The final consequence is that the central government has to reconsider its direct intervention in local governance, which in turn makes local authorities get into the troubles. For instance, the central government has empowered right to the municipal government and encouraged it to carry out a finance innovation so as to address its financial shortfalls, while on opposite it has to regulate the irregularities of local authorities by limiting its behaviors.

Chapter 5 Industrial Land Redevelopment and Profit-Driven Relationships

5.1 The initiation of industrial relocation and land re-mediation

5.1.1 Placing relocation on the agenda—dilemma and opportunities

As described in the previous chapter, the high-efficient use of urban land has been advocated by Lanzhou's municipal government as a strategy to solve the urban problems occurring since the 2000s, especially after the central state's strict limitation on the land conversion from agricultural uses into construction purposes in 2007. In the early period, the reuse of inefficient-use lands in the municipal agenda referred primarily to the redevelopment of residential and commercial lands in inner city, such as the transformation movement in the urban villages and shanty towns around 2005. The municipal plan did not involve the redevelopment of industrial land at beginning.

One major reason is that after the long-term lack of construction and maintenance of housing and living facilities (since the planned economy period), Lanzhou municipal government has realized that the improvement of the dilapidated residential and commercial spaces are more urgent on its agenda of the city's image and livelihood projects. Meanwhile, the existing urban villages and shanty towns were considered to provide a sufficient quantity of land to local land market, thereby contributing to the municipal finance. Moreover, in consideration of the difficulties in industrial demolition and relocation (such as high relocation costs, new production siting, the placement of workers, and so on), the contribution of the industrial sector to local GDP and employment, and the stable relationships with individual enterprises, Lanzhou municipal government was not keen on the industrial land redevelopment.

After the nearly five years' redevelopment of urban villages and shanty towns, the Lanzhou municipal government faced a serious land shortage in 2010. On the one hand, the fast population growth in urban areas and the consequent land demands for infrastructure asked the municipal government to look for new land resources.

It is estimated that the stock of construction land in Lanzhou city remains less than 1 million acres, in which more than half is the inefficient-used industrial lands. With the annual supply of 2,000 acres since 2007, the remaining construction land is only available for the city's 5 year of construction demand — an interview with Li, the deputy director of Lanzhou Land Resources Bureau in August 2013[1].

On the other hand, the dwindling and stretched municipal finances forced the municipality in doing so. In many Chinese cities, the land-related revenue has on average accounted for 30% to 40% of local fiscal incomes.

Yet in Lanzhou, this proportion has been less than 20% because of its valley topography. In 2011, the land transfer income in Lanzhou accounted only for around 15% of local

---

1   Source: form the website: http://www.nbd.com.cn/articles/2013-08-01/762965.html.

finances. Although the municipal government has looked for other ways to increase local revenues, such as the introduction of industries and inward investment, it seems ineffective. Such financial situation made the provincial and municipal leaders rather embarrassed—(ibid.).

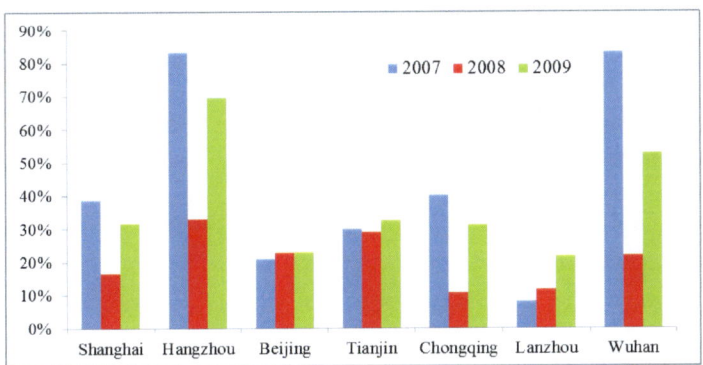

Source: data of land revenue are collected from the Land Transfer Survey Report of China Index Academy (2009), and data of local fiscal revenue are collect from the Statistical Yearbooks (2010) of each city.

Figure 5. 1 Share of land revenue in municipal fiscal revenues of Lanzhou

Obviously, the shortage of land supply has stimulated a rapid growth in commercial and residential land prices. By comparing the land benchmark premium[2]—i.e. the minimum price of land

_____

2   A benchmark land price system provides guidelines for land use rights selling and transferring. It was published to overcome the lack of market data and experiences in land transaction. It provides the minimum price of land transactions at different locations which are separated into seven levels according

transactions—of Lanzhou, it can be found that the commercial and residential land near the city center (level 1-2) has raised an average of 1000 yuan within a decade (Figure 5.2). In contrast, the basic transaction price of industrial land remains low.

The land transaction price determines the municipal land-related incomes. If the municipal government acquires the industrial land from enterprises at the benchmark price of industrial land (or by paying a slightly higher price than the industrial transfer fees as compensations) and then transfers the lands in the name of residential and commercial purposes, the resulting profits are rather substantial.

> The municipal government recovered a piece of industrial land (with an area of 30,000 m²) in Chengguan District in 2008. The land was initially planned for an urban park and cultural plaza, but the project was shelved for various reasons. Three years later, the government decided to transfer part of the land for commercial and residential uses for a consideration of its prime location. The final transaction price was about 8,000 yuan/m², which is almost double of that of three years ago— interview with a Lanzhou Commissioner.

---

to the distance from the city center. The benchmark prices of land use rights are determined by land use, land use density (floor-land ratio), land grades, land improvement, and tenant resettlement costs..

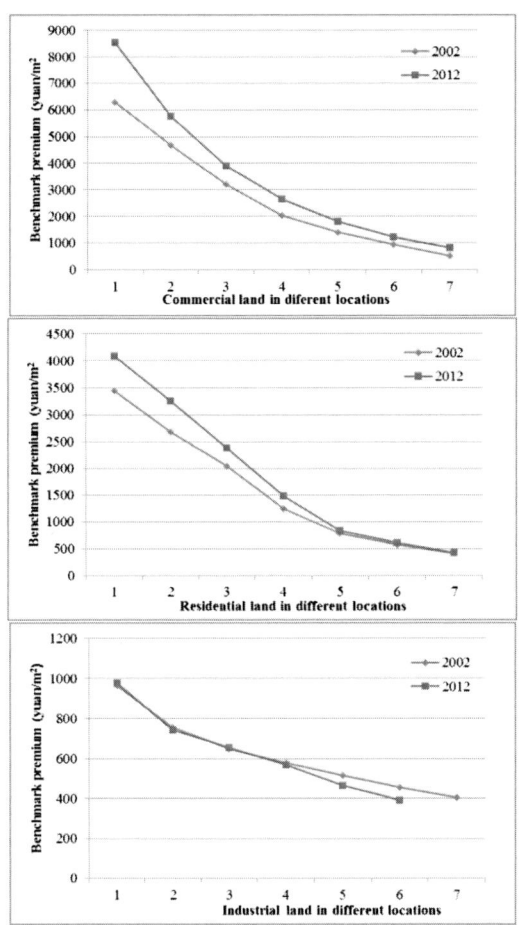

Source: drawn by the author based on the data from Lanzhou Land &Resources Bureau.

Figure 5. 2 Benchmark land prices[3] of Lanzhou city in 2002 and 2012

---

3   Ibid.

For multiple considerations, the redevelopment of industrial land has become the focus of the urban construction and livelihood projects of the Lanzhou municipal government since 2010.

The municipal government's dilemma — being limited autonomy and disintegrated authority

Even though Lanzhou municipal government has been well aware of the huge potential of land appreciation, it is not easy for it to implement the renewal project. The main obstacles come from the fuzzy land ownership and the ambiguous administrative relations with local enterprises. In most of the old industrial areas, the property structure of enterprises is rather complex. For instance, in the Xigu district of Lanzhou city, there were around 200 industrial enterprises by the end of 2010, seven of them are state-controlled enterprises that three of which belong to the military, ten are provincial-controlled enterprises, seventeen are municipal-owned enterprises, and the remaining 100 enterprises are private-owned[4]. If the municipal government wishes to withdraw the industrial land from the central- and provincial- controlled enterprises, normally it needs the approvals of the central ministries and the Central Military Commission. Even if the management of urban land ownerships has been entrusted by the central state to the municipal government and the local authorities also have legitimacy in doing so, the autonomy at the local level is constrained within the hierarchical regime.

The rationality and feasibility of Lanzhou Petrochemical Company's (the subsidiary of China National Petroleum

---

4   SOEs in the article refers to the government-owned enterprises, government-holding enterprises and government shareholding enterprises)

Corporation and biggest central-owned enterprise in Lanzhou) relocation need to be comprehensively argued. "Move or not to move" cannot be decided by the company itself as well as the municipal government. Its relocation should be considered from the perspective of the "oil strategic layout of the country"—an interview with Zhang, a professor of Lanzhou University, reproduced from The Paper News[5].

In addition to this potential influence from the central government, the authority of local states that was once unchallengeable is also threatened. The diversification of business property implies that local officials can no longer issue the forced commands to most of the enterprises. In face of the renewal project, these enterprises often hold different attitudes towards the relocation and have diverse expectations regarding issues of compensation and placement. The central- and provincial-owned enterprises, by virtue of their close relationship with the central ministries and the military, normally have strong voices in the negotiation over land acquisition and redevelopment. For the private enterprises, it is also impossible for the municipal government to withdraw the land from them. After the "Regulations on the Assignment and Transfer of the Land-use Right of the State-owned Land in Urban Areas" was published by the central state in 1988, the vast majority of private industrial enterprises have signed the land transfer contract (50-years to use) with Lanzhou municipal government. This means the way of that the municipality withdraw the land use rights before they have expired has lost its legitimacy.

---

5   Quote from: http://www.thepaper.cn/baidu.jsp?contid=1297557.

Making the relocation rationalized
However, it is difficult but not impossible for the municipal government to recover the land-use rights. In order to make the industrial relocation and land redevelopment rationalized, the municipal government has been devoted to finding reasonable grounds.

According to the relevant laws and national regulations[6], the land use rights could be withdrawn by municipal governments in the following cases. For lands involved in the urban renewal programs and public projects in the latest urban master planning, the municipal governments have the right to withdraw them; and for those being allocated to the land users in the planned period after the enterprises' relocation, bankruptcy, or the termination of the land-use rights for other reasons, municipal governments are permitted to recover the lands. Based on these two points, the municipal government has promulgated four planning, namely the Fourth Edition of Lanzhou City Master Plan (2011-2020), Lanzhou Land Remediation Planning (2011-2015), and Lanzhou Land Use Planning (published in 2012). First, Lanzhou municipal government commissioned the Lanzhou Municipal Planning Bureau and China Academy of Urban Planning and Design (Beijing) to re-adjust the land-use structure in the city center and reduced the planned proportion of industrial lands in the total area of urban construction lands (from the current 38.15% in 2012 to less

---

6   In the People's Republic of China Land Management Law (1999), Article 58; the Urban Real Estate Administration Law of People's Republic of China (1995), Article 19, and the Regulations on the Assignment and Transfer of the Land-use Right of the State-owned Land in Urban Areas (1988 and 2006), Article 42 and 47.

than 10% in 2020)[7] in the master plan. Except for a small amount of industrial land located in the urban fringes, the vast majority of existing industrial lands in urban areas are planned for the residential and commercial purposes (Figure5.3).

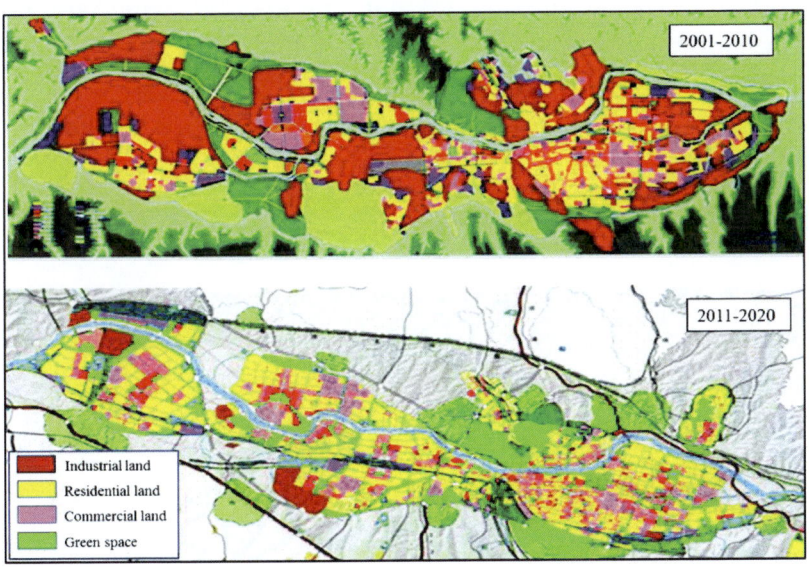

Source: from the official website of Lanzhou City Planning and Design Institute

Figure 5. 3 Lanzhou City Master Plans (2001-2010 and 2011-2020)

At the same time, the municipal government proposed an intensive and efficient goal for the land development in the land remediation planning and land-use planning. According to the Urban Land

---

7   Data are collected from Lanzhou City Master Planning (2011-2020) and Lanzhou Land Use Planning (published in 2012).

Classification and the Land Construction Standard of Planning introduced by the central state in 2012, the proportion of industrial land in the total area of construction lands has been strictly limited between 15% and 25%. In Lanzhou, however, the proportion (38.15% in 2012) is much higher than the national average, especially in its two traditional industrial districts—the Qilihe District (around 36% in 2009) and the Xigu District (around 44% in 2009)[8]. This gives the municipal government a good excuse to re-enact the standards of industrial land.

In its planning, the municipal government pointed out:

> Lanzhou is currently facing 'serious' land issues, such as the inefficient use of land and the contradiction between land supply and demand...land remediation has risen to a 'national strategic level' and become a 'vital' method to maintain economic development, to protect food security, and to promote the people's livelihood...thus, it is necessary to determine a 'reasonable' size and growth rate of industrial land in the city and to improve the level of land intensive use—extracted from Lanzhou Land Remediation Planning (2011-2015).

By estimating the potential supply of land after the urban renewal project, Lanzhou municipal government reached a conclusion. "The newly increased 44.9 million m² of land in which the industrial sites are expected to account for 59.5% of the total supply

---

8   Source from: the Overall Land-use Planning of the Xigu district (2009-2020)ö downloaded from the website:
http://www.gsdlr.gov.cn/lzxgq/content.aspx?id=ARTI98

area, will support around 60 million new urban residents and improve the city's living environment" (Lanzhou Land Remediation Planning '2011-2015').

5.1.2 Starting the redevelopment program—seeking for assistance from superior governments

As early as 2006, Lanzhou municipal government has proposed a relocation plan named Industrial Enterprises' Relocation and the Transformation of Industrial Districts. This plan was not put into action until the municipal government issued the city master planning (2011-2020) and supporting land planning in 2010 and 2011, which provide a legal basis for the municipal relocation plan. To supplement the relocation plan, the municipality issued a detailed program entitled the 12th Five-year Plan for Enterprises Relocation and the Transformation of Industrial Districts (2011-2015). In this program, a total number of 157 industrial enterprises in the urban area were included in the relocation agenda. This amount accounted for 2.33% of all industrial enterprises of the city in 2010. Most of them, according to the official statements, are the heavy-polluting enterprises or those occupying excessive amount of land in the urban core area or whose sites have been planned for the municipal infrastructure (e.g. urban main road, subway, and high-speed railway station) in its earlier master plan.

The enterprises on the relocation list are involved in the businesses of machinery manufacturing (38.85%), building materials production and processing (24.20%), chemical manufacturing (11.46%), textiles (6.4%), processing of agricultural products (5.10%) and biomedical manufacturing (3.2%), and relate to the state-owned, joint-stock, as well as private sectors...after the relocation, these lands will be developed for commercial and residential uses...except under special circumstances, land development for the industrial purpose on the original sites will not be approved—extracted from 12th Five-year Plan for Enterprises Relocation and the Transformation of Industrial Districts (2011-2015).

Despite having introduced a detailed relocation program with clear objectives and this program seems to be reasonable and has its legal basis, Lanzhou municipal government was not able to carry out its action immediately. As mentioned before, many of them listed on the relocation agenda are central- and provincial-owned enterprises which were founded in the planned economy era. Given their insurmountable administrative hierarchy, local authorities have no right to intervene in the decision-making of these enterprises, let along forcing them to relocate. In consideration of their long-term stable relationship in the past, local authorities are also not willing to break this this balance. For the municipal government, it is also not easy to ask the private enterprises to relocate. Comparing with the central- and provincial enterprise, the bargaining power of the municipal and private enterprises are relatively weak. However, encouraged by the tough attitude of the central- and provincial-owned enterprises towards the municipality, they simultaneously have excuses to refuse the relocation.

In addition, another challenge confronted by the municipal government is where the enterprises should relocate to. This is often used by enterprises as an excuse for refusing to move. Between the late 1990s and 2010, Lanzhou municipal government has already established four industrial development zones at the urban fringe[9]. The main purpose of their constructions at beginning is to attract foreign investment in Lanzhou. As tightly surrounded by residential and commercial buildings as well as supporting facilities with the rapid of urbanization, the remaining lands in and surrounding these industrial zones are insufficient to provide more spaces for such massive relocations. Nevertheless, finding a reasonable solution for this thorny issue is not completely out of reach for the municipal government.

In 2011, Lanzhou municipal government launched a strategy, entitled "To Build a New Urban District outside the Valley Basin" ("跳出老城、建设新区，跨越发展、再造兰州"). In this strategy, the government proposed that a new industrial district would be built at the same time as the inner city renewal. Except for providing space for the industrial relocation, another ambitious goal of the government is to make the area of urban construction land doubled in the next 15 years [10]. After reporting the strategic conception to its superior—the Gansu provincial government and obtaining approval, Lanzhou municipal government commissioned the provincial government to issue a policy at the provincial level, named the Provincial Guidance on the Construction of Lanzhou New Industrial District. In this policy, Qinwangchuan town, belonging to Lanzhou Gaolan County and

9   Lanzhou High-tech Development Zone in Qilihe District, Lanzhou Economic and Technological Development Zone in Anning District, Xigu Industrial Development Park in Xigu District, and Heping Industrial & Development Park in Yuzhong County.

10   Data is from Lanzhou Television Network, http://www.lzr.com.cn/view/view_14088.html.

located 35 km away from the main city of Lanzhou was designated as a concentrated area (around 806 km²) with an aim to attract 80% of industrial enterprises moving out from the inner city and those from neighboring cities and provinces (Figure 5.4).

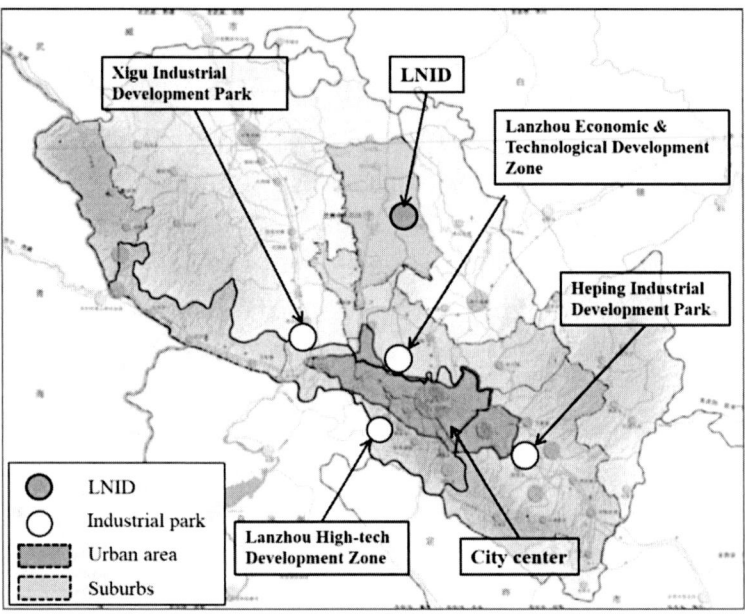

Source: drawn by the author based on the Lanzhou administrative map.

Figure 5. 4 Main city and industrial development zones of Lanzhou administrative region

In accordance with the government's forecast, the establishment of Lanzhou New Industrial District (LNID) will play a certain role in promoting the enterprises relocation in inner city. At least the lack of new production site is no longer a reasonable excuse for enterprises. Meanwhile, the administrative committee of LNID authorized by the municipal government promised publicly to provide the relocated enterprises with preferential policies regarding new production siting, tax deductions, infrastructure, ancillary services, and others issues. However, the progress of relocation is still far below the government's expectation. By the end of 2011, fewer than 10 enterprises moved to the LNID. In face of the repeated mobilization of provincial and municipal governments, many of them showed great reluctance on the relocation and took various countermeasures to delay their moving. A typical example is the subsidiary of China National Petroleum Corporation (CNPC)—Lanzhou Petrochemical Company (LPC) which is the largest central-controlled enterprises located in the Xigu district. The relocation cost of this company is estimated about 60 billion yuan. Its business leaders claimed that the relocation would result in a resettlement of nearly 10,000 employees. The company cannot disregard the strong opposition from its employees and afford the huge costs of relocation and staff resettlement alone. Such a huge relocation project and the resulting compensation fees make many large enterprises discouraged. The vast majority of enterprises on the relocation list tended to take a wait-and-see attitude considering the high relocation costs and difficulties of staff placement and meanwhile having expectations of more preferential terms from the municipal government. Besides, they are greatly skeptical of the development prospects of the LNID due to its far-off location.

However, the stalemate situation has somewhat reversed since 2011. First, the State Council issued the National 12th Five-year Plan for Environmental Protection in 2011 in response to the country's serious environmental issues. Environmental protection becomes one of the important indicators in the performance evaluation of local officials. For Lanzhou municipal government, however, a greater significance of this plan is that it is authorized to urge the relocation of the polluted enterprises in the name of environmental governance. This also includes those arrogant central- and provincial-state enterprises.

At the same year, Lanzhou municipal government jointly with Gansu provincial government applied for the upgrading of the LNID as the Fifth State-level Industrial District (FSID). The upgrade of LNID to the state-level assists the municipal government in obtaining national supports on policy formulation and implementation. In the official documents issued by the State council, respectively the Approval of State Council on the establishment of FSID and the National planning of the transformation of the old industrial base (2013-2022), it is required that all sectors of society should spare no efforts in supporting the construction of LNID (or FSID) and the transformation of Lanzhou old industrial districts. Certainly, the SOEs as well as their superior administrative authorities are also covered. In addition, Lanzhou municipal government put pressure on the enterprises by means of public opinion since 2014. It is the first time for the government to openly criticize the polluting enterprises through the social media in the name of environmental protection and public health.

After the long tug of war, the vast majority of enterprises on the relocation list have eventually signed the relocation agreements with Lanzhou municipal government either willingly or unwillingly. By the end of 2014, around 100 industrial enterprises on the list have finished relocation or have agreed on the relocation issues with Lanzhou municipal government, which provide nearly

8 million m² of industrial land for commercial and residential development.

5.2 The restructured interest-relations: a profit-driven target and its divergences

The redevelopment of the old industrial district involves land acquisition, land transactions, capital investment, and project construction as well as operation. These processes will inevitably break the balance of interests which has formed on the basis of existing property rights, and rebuild the interest contacts among stakeholders. Meanwhile, the dualistic land ownership, regulated and non-regulated land market, chaotic financing ways, increasingly blurred administrative hierarchy, and so on will make these relationships further complicated and diverse than those in the planned economic era.

On the one hand, enterprises are encouraged by the mobilization and simultaneously coerced by the strict punishments from the central and municipal levels. On the other, some of them are tempted by the booming real-estate market and have sprouted out of a desire for speculative capital development. Meanwhile, private developers and broad social capitals are also involved in the urban construction. In line with the central and municipal governments' vision, they are going to assume a primary role in urban renewal and land redevelopment. Within the framework of government-dominated urban renewal, governments, relocated enterprises, and developers have reached a tacit understanding—i.e., the pursuit of profit in land development. In consideration of maximizing self-interest, however, they have various divergences on resource and responsibility allocations, risk taking, and profit distribution. It thereby triggers various conflicts between stakeholders. As such, consensus and conflicts are co-existed in the renewal process.

5.2.1 A profit-driven target of stakeholders

The demands of enterprises

Although the majority of enterprises in the last ten years did not give enthusiastic and positive responses as Lanzhou municipal government expected, it does not mean that they have no motivation to relocate. In fact, many of them showed great interest. The two main purposes of their reluctant relocation, as mentioned before, are the concerns over the high relocation costs and the uncertain prospects for LNID, and the high expectations of more preferential policies.

Since the mid-1990s, when the land value in the inner city increases with the rapid growth of the real estate market year after year, many enterprises have taken note of this. In particular, those with poor economic performance and facing bankruptcy realized: the profits enterprises created through the production are probably much lower than the income generated by the transfer of land, or than that raised from the sale and lease of building and facilities attaching on the ground after a property development (e.g. commercial housing, office building, hotels, and convention and exhibition facilities). For instance, Lanzhou Textile Co., Ltd. had a loss of 4.22 million yuan in 2008, while it profited 11.9 million yuan in 2010 through transferring 6,000 m² of land on the original site[11]. Faced with such huge returns, some cannot wait to sell their land

11   Date from Gansu Xinhuanet, website:
http://www.gs.xinhuanet.com/news/2005-11/09/content_5547274.htm

in the city center, or to change their business and engage in the commercial and residential redevelopment on their production sites.

In view of the inadequate construction land in the inner city, some enterprises are unable to expand the workshops and production facilities in place to meet the demands of their production expansion. It seems to be a sensible and inevitable choice for them to escape the city center.

> The provincial and municipal governments have provided the unprecedented preferential policies and financial subsidies. In addition to tax relief and the provision of new production sites with low rents, the municipality also allows the factories to redevelop the land by themselves in its latest policies. Undoubtedly, it is the most appropriate time to relocate — a dictation of a serving staff, reproduced from Tianya Forum, the community of Lanzhou Petroleum Machinery Factory.

Meanwhile, the continuous growth of land rent in inner city has invisibly raised the production costs of enterprises. If enterprises move into the suburbs, they may probably obtain a sufficient amount of production land as well as cheaper rental prices from the municipal government or the committee of industrial parks. Moreover, the favorable infrastructure and service environment in the industrial zones and the "spillover effect" resulting from the agglomeration of similar businesses strongly appeal to them.

In addition, some enterprises have received fine from the local environmental protection department frequently in recent years because of the repeated pollution and explosion incidents. LPC had received several fines due to the previous air and water pollution as well as gas explosions. The latest punishment was for the city water pollution incident in 2014. The company therefore paid 100 million yuan in fines. The penalty was used to fund the purification of groundwater and the soil remediation.

> The series of actions of the municipal governments are obviously against us. For example, it asked the Environmental Protection Bureau to come forward and criticized us in the public media. The mayor excused that he was unaware of the matter, and explained that LPC's relocation is good for environmental protection and the public. But we all know that government doing so is to force us to move out. They hope to flip the un-optimistic development trend in Lanzhou New Industrial District (LNID) through the hundreds of billions of relocation project. We also want to move and contribute to the city's environmental protection and public health. Yet we indeed have difficulties—a LPC's executives, reproduced from Beijing Times[12].

Under pressure from increasingly stringent environmental emission standards and the lack of funds for renovation of aging facilities and equipment, many polluting enterprises are caught in a dilemma. Faced with the special financial subsidies of government-sponsored program (named "relocation and equipment upgrades"), they are inevitably not tempted. Similarly, for those companies which have years of losses, it would be an opportunity for a turnaround.

---

12   Quote from http://epaper.jinghua.cn/html/2015-01/12/content_161460.htm.

The demands of private developers

The most fundamental motivation of developers and investors involving the transformation of the old industrial areas is to maximize their capital gains through the investment in land, infrastructure, and buildings. More importantly, they hope to create an urban physical environment in anticipation of capital appreciation. The issue they are most concerned about is the cost, in particularly the upfront investment in the land purchase and infrastructure construction. Whether they are able to access lands at low enough prices and thereby reap benefits is the most fundamental reason for their decision to invest. Compared to the prices of commercial, office, and residential land in urban areas, the average land price in old industrial areas is relatively inexpensive; even if these lands are auctioned in accordance with the benchmark premium of commercial and residential land after land-property conversions. In their views, these plots have great potential for appreciation after a steady stream of government financial supports and social investment, and the resulting improvement of the surrounding physical environment in the future.

Compared to repeatedly shoot sky-high price of commercial, office and residential land, industrial lands are inexpensive and with great potential for appreciation. There is no harm to take more. We can "attack and defend" in the unpredictable market—a dictation of a consultant of the real-estate development of E-Power Group[13].

---

13   Quote from: http://www.gse-power.com/W/HdContentDisp-5-1503-2009612-199162.htm.

Normally, infrastructure on the industrial land (e.g. road network, power grid, water supply, etc.) has been well established. It helps to reduce the costs for developers, freeing fund that will be spent in their land primary development. Moreover, unlike the projects involving commercial and residential area redevelopment, in which developers are normally asked to be in charge of most of the relocation compensation for the households and tenants, these are completely unnecessary for them to take into account in the cases of industrial redevelopment. Additionally, as both central and local governments pay special attention to the urban renewal programs, they tend to offer more favorable policies compared to the other municipal projects so as to attract a broader range of investors.

> For the "three old renewal" project, in general, the taxes will be lower than that of the ordinary real-estate projects. This is sufficient attractive—a dictation of a real estate manager, reproduced from Southern Metropolis Daily[14].

Based on these reasons, the redevelopment of industrial land is highly favored by developers.

5.2.2 The divergence of interests
In view of the potential profit that urban renewal and land redevelopment would probably bring about, the expectations of project returns for the interest groups are widely divergent. Given their divergence, the competitions over the division of responsibilities and benefits between stakeholders continue.

---

14   Quote from: http://epaper.oeeee.com/epaper/L/html/2014-08/22/content_3300496.htm?div=4.

Municipal governments and enterprises

The conflicts between governments and enterprises are primarily centered around the debates over relocation compensation (i.e. equipment impairment charges, lost income, land compensation paid to the original land users of new sites, and resettlement costs of employees), land ownership, development rights of the former sites, as well as the profit distribution after the land transfer/redevelopment.

From the enterprises' side, they prefer government bearing most of the relocation costs. However, governments only hope to offer non-financial subsidies, such as preferential land policies, tax breaks, and the assistance on the personnel placement and employee benefits (e.g. unemployment insurance, re-employment and low-rent housing for workers who live in difficult). Especially for those with poor economic performance, governments wish they could solve the dilemma by seeking financial support from those in good financial situation, after being merged or acquired. However, this is not accepted by the majority of enterprises since they do not want to give up operating autonomy. Additionally, taking into account of local financial pressures and political achievements, local officials hope that industrial land in the city center would be vacated and built for other uses as soon as possible. In opposite, enterprises are rather cautious about the land transaction whether transferring their lands to governments or in the market. In view of land prices skyrocketing since 2004, the corporate leaders are convinced that the later they move out from the city center, the more land revenue they could earn. More importantly, a higher value of the land can increase their bargaining power in the negotiations with local government.

Not only our factory (Lanzhou Yellow River Brewery Co., Ltd.) is unwilling to move, other factories in this area are not yet plan to relocate. This area, for its good location (adjacent to the Yellow River), has been planned by the Municipal Planning Bureau as a sport center, a convention center, and several commercial and residential buildings. Land and housing prices continue to rise every year. The latter we move, the more valuable our lands will be. Our factory as well as several others has reached a tacit understanding. The municipality is more anxious, but not us—an interview with a retired worker of Lanzhou Yellow River Brewery Co., Ltd. in October 2013.

Another debate between enterprises and the municipal government is around the issue of land use conversion. Normally, enterprises hope to keep the land use-right after their relocation. In this way, they only need to pay the municipal government the low rents by reference to the transfer standard of industrial land.

If the enterprises can continue to hold their industrial land property, they could transfer the development and management rights secretly to private developers through the intermediaries and make profits, such as the building facilities facing the street. This is permitted if enterprises build facilities and operate them for the commercial uses by itself. The secret transactions are rather common since the late 1990s. The municipal and district governments are turning a blind eye to it. So enterprises certainly do not hope to return their lands to governments, or to change the usage of the plot—an interview with a worker of Lanzhou Build Power Factory in October 2013.

Subject to the regulation on the rental price of industrial land, savvy owners of enterprises usually lease the idle lands or premises at a rather low price to the "principal tenants" with whom the owners keep a close relationship. After that, these principal tenants sublease the lands to the users at the prices of commercial and residential land. In this way, enterprises can not only evade the liability for illegal alterations but also gain huge profits resulting from the price differentials. It thus brings about a huge "gray secondary market" and a multitude of illegal constructions. After the municipal government launched its industrial relocation program in 2010, this approach has been strictly forbidden. In the name of market supervision, the municipal government has asked them to recover the land from the tenants or change land usage into commercial or residential usages.

Ordering the enterprises to reclaim the land and building from their illegal tenant is just the first step of governments. The second step would be to ask these enterprises to hand over their land to government or transfer the land in the public— (ibid.).

The third contradiction is that both Lanzhou municipal government and enterprises hope to occupy the land-use rights after the relocation. If conditions allow, enterprises desire to formulate the reconstruction projects on their own and redevelop the lands according to their intentions. As thus, enterprises can access the vast majority of profits through land and housing appreciation. After once paying off the land-related taxes to the municipal governments, they occupy the residual incomes, including the incomes from real estate leasing and selling, and the potential interests which will be generated from land appreciation.

In contrast, the municipal government is more willing to recover the lands by refunding the remaining land-rents that enterprises have paid before. In this way, governments can access a more considerable amount of returns after the land auction, not only including the taxes that enterprises turned over.

Compared to the net transfer of land, the sale and lease of the developed land as well as the attached facilities on the ground are apparently able to bring enterprises more substantial revenue. For instance, the average transaction price of commercial buildings and retail stores (82,000 yuan/m²) in the Chengguan district[15] is about ten times as much as its benchmark land price (8,540 yuan/m²) in 2012[16].

> In the core areas of the Chengguan and Qilihe districts, as long as enterprises invest and build the commercial and residential facilities and sell them at an average price of local real estate, most of them are definitely the winners—an interview with a Lanzhou Land Commissioner.

The fourth controversial issue involves the specific projects regarding the redevelopment. Enterprises that have the potential to engage in an independent redevelopment[17] normally desire to

---

15   Data are collected from the "Report on New Commercial housing and Retail Stores in Lanzhou Urban Areas between January 2012 and June 2013".

16   Data are collected from "Lanzhou Benchmark Price of Commercial Land".

17   Before 2010, they are mainly those making significant contributions to local GDP, or having maintained close relationships with local officials, such as the large-scale SOEs; while between 2010 and 2014, any enterprises are allowed to carry out the independent land redevelopment.

develop commercial housing and commercial office buildings as much as possible. Given public interests, the municipal governments sometimes require them to transfer part of industrial lands for public projects, such as squares, green spaces, schools, hospitals, railway stations, and so on. According to the Management Approach to the State-owned Land Transfer Revenue and Expenditure of Gansu Province, the price of compensation that Lanzhou municipal government pays to the existing land users for the public projects is normally far below the average price of commercial and residential land in the same area. Given the forced enforceability of urban planning, enterprises often cannot reject these projects. For instance, Gansu Province Chinese Medicine Hospital submitted an application of land acquisition to Gansu Provincial Health Department and Lanzhou municipal government in 2009. Under the coordination of Lanzhou municipal government, the hospital obtained 14,708.74 m² of land from Lanzhou Changsheng Vegetable Oil Co., Ltd. for a purpose of expansion. Although the company repeatedly demanded for increasing compensation, the hospital paid a small portion of the amount that the company claimed. This amount is only equivalent to the cost of demolition and resettlement. Even though the compensation is in line with the official standard, as a mediator, the municipal government eventually has to promise the company to provide a new production site at an inexpensive price.

Municipal governments and private developers
The conflicts between local governments and developers are mainly around the quantity and price of land in the market, as well as the contents and construction standards of redevelopment projects. By setting up the preconditions of land purchase and

project development, governments guide developers' behavior and force them to perform certain duties in the renewal process. In addition to the desire of getting higher land bid prices, governments also expect that developers will assume some construction tasks (mainly public construction projects) and obtain their own pre-construction funds as much as possible. However, in order to get more favorable terms and maximize their own interests, developers continuously try to break the regulations and permissions formulated by local governments or to evade their obligations in the agreements having been reached between them.

First, local governments would like to tighten the total supply of land on the market by virtue of the land storage system. It helps to raise the price of land at auction by triggering fierce competitions between developers. Even though the high land acquisition costs will eventually be passed onto consumers, the inadequate land supply is in fact inconsistent with the goal of developers (the pursuit of short-term profit). The shortage of land supply makes real-estate companies "landless." As a result, they have to hoard the existing land to ensure profit growth which in turn leads to more severe land scarcity and higher land prices. However, for the urban renewal project, this is forbidden. Real estate developers are normally asked them to develop immediately once they access the land.

We can say that the "three old renewal" projects are tailored for large developers, with which middle and small developers must be very prudent to treat. Some renewal projects even involve the issue of unclear property rights (i.e. shantytowns), which makes small developers be more cautious. Thus, in order to attract developers to participate, local governments have to deliver the projects to developers' "home" and please them to take the projects—a dictation of Guo, a manager of Centaline Property Company[18].

Second, governments are solely in charge of the contents and construction standards regarding the renewal projects. They wish these projects to be consistent with the existing planning and regulations as much as possible.

Developers certainly hope that the price of land is as low as possible; the more profitable projects are, the better it is. If the government cannot provide, the developers have to find a "way"[19].

After the land auction, governments normally require the developers to assume several public welfare projects or invest the necessary infrastructure in renewal projects. To increase profits, developers sometimes raise the volume rate and building density or change the land use types without a formally official authorization. However in some cases, such act of developers is sometimes agreed privately by local officials before the auction.

Enterprises and private developers

The conflict of interests between relocated enterprises and developers focus primarily on project redevelopment, while in terms of land transaction and enterprise relocation, this is less so. In order to avoid the irregular transaction behaviors and protect

---

18   Quote from: http://epaper.oeeee.com/epaper/L/html/2014-08/22/content_3300496.htm?div=4.

19   Ibid.

their own interests, Lanzhou municipal government strictly prohibited enterprises looking for land traders themselves, which was once encouraged by the municipality before 2005. Since 2006, the vast majority of land transactions are required to be tendered and auctioned on the public platforms built by local governments. After being approved by the municipalities, the relocated enterprises are able to sign contracts with developers under the supervision of the relevant administrative departments. The establishment of the land reserve and transfer platforms (e.g. LLRIC) has cut off the illegal interest-chains between enterprises and developers, which is formed during the transactions of lands and ground attachments, such as the black-box operations on land transfer price and the illegal profit-sharing between local officials, land users, and land applicant. After the introduction of public auction, land transactions only involve land-use rights and take the benchmark price as a reference. In this case, there are few conflicts between enterprises and developers.

Since 2008, Lanzhou municipal government permitted the relocated enterprises to engage in the redevelopment of their former industrial land. They are allowed to conduct the real-estate development by themselves or with the assistance of private developers. In the latter case, the conflicts of interest between enterprises and developers focus primarily on the pre-investment and the distribution of profits. Normally, enterprises pledge to provide land and a small amount of construction funds and require developers to be in charge of the initial investments, project applications, and specific construction. They desire the developers to offer funds for the projects as much as possible.

The potential bidders should be those in good financial condition and having a good bank credit rating and the ability to pay off the entire land transfer fees immediately. Besides, the projects and construction standards must meet the requirements of the Overall Redevelopment Planning of the LLG's Original Site, which has been approved by the Lanzhou Municipal Planning Bureau and municipal government — excerpted from the Project Tenders of the Commercial Real-estate Development of LLG.

Although Lanzhou Petroleum Machinery Factory has published a highly tender price and bidding conditions of its several plots, the developers coming to bid are in an endless stream. This location is too tempting. If conditions are not too harsh, there should be more bidders — a dictation of a serving staff, reproduced from Tianya Forum, the community of Lanzhou Petroleum Machinery Factory.

In addition to the profitable projects, enterprises also hope the developers will assume some welfare projects (such as financing housing for their employees) or help the enterprises to solve the employment problem of workers. Except for the commodity housing, office buildings, hotels, and commercial and trade centers, developers are sometimes required to build the enterprises' R&D department and a number of employees housing. These building projects are usually transferred by the developers at a cheap price (normally less than half of the market price) to the enterprises. Then, employees acquire the housing by paying slightly higher prices than that their companies have paid. However, they only access to partial ownership, half of which is held by their companies. These housing can neither be sold nor be rent out in market, but are allowed to be transacted among the employees.

There is another case. After the completion of the commercial projects jointly developed by enterprises and developers, as exchange, enterprises normally transfer the newly-built facilities to the developers at relatively favorable prices. Or sometimes, enterprises also sell or lease the commercial facilities to other users, or use these facilities to expand their business. In the former, the developers and new tenants are asked to hire the staffs of the enterprises that are unwilling to migrate with their companies or to provide jobs to their families as compensation for the relocation. This is normally a premise that the new tenants or developers have access to the management rights of land and commercial facilities. For instance, Lanzhou Iron & Steel Factory, a state-owned enterprise covering an area of 533,000 m² in Chengguan district, was redeveloped as the largest logistics center in the city after the bankruptcy and the mergers and acquisitions by Jiuquan Iron and Steel Co., Ltd in 2002. After then, all plant and equipment of the former factory were relocated into Yuzhong County, 30 km far away from Lanzhou city center. The original industrial land was redeveloped for a logistic center by the Jiuquan Iron and Steel Co., Ltd. After the completion, the logistic center hired 1,500 new employees, in which 400 positions were asked to provide to the unemployed staffs that are caused by the bankruptcy and relocation.

Nevertheless, enterprises are not always dominant. Sometimes they have to compromise with developers on the project contents as well as profit-sharing, owing to their reliance on the funds and development experience from developers. This is very common for underperforming enterprises and the ones that are in disadvantaged locations. In order to cut the high costs caused from the involvement of developer, more and more enterprises start to consider establishing their own real-estate subsidiaries. According to the fieldworks in September and October 2013, two of 14 interviewed enterprises have set up their own real estate subsidiaries and have the experience of engaging in real estate development projects. One of the enterprises, CNR Lanzhou Locomotive Co., Ltd. plans to set up its own real estate subsidiary in the next few years. The three companies are large state-owned enterprises with strong financial backgrounds. Six of 14 enterprises, being incapable of setting up their own real-estate departments, have already conducted or are going to engage in the redevelopment with the assistance of third-party developers. The other five enterprises chose to transfer the land directly through the LLRIC after relocation, and they are no longer involved in any projects related to the redevelopment of the original industrial sites.

5.3 Policy incentives for the redevelopment

5.3.1 Making concessions

The municipal officials urgently desire a considerable amount of fiscal revenue and are more importantly eager to stand out in the competition of political achievements during their brief tenure. However, it seems to be unrealistic to achieve the goals, fully ignoring the demands from developers and enterprises. In order to accelerate the pace of reconstruction, the municipal government has to sometimes make a little concession, such as transferring part of the interests to other stakeholders. Thus, it realizes that it is necessary to develop a policy framework in favor of multilateral cooperation, and kept looking for the conflicts and formulated specific policies in response to the issues that hinder the smooth implementation of its renewal projects.

Lanzhou municipal government is a legal representative engaging in organizing local economic production and managing public affairs in the city. On Behalf of the central state, it monopolized the urban land ownerships, and at the same time it has authority on local tax administration, urban planning, policy-making, and coordinating the actions of administrative institutions, organizations and individuals. Thus, the municipal government has strong ability to re-distribute the interests by formal or informal means.

First, Lanzhou municipal government has a monopoly on drawing up, examining and adjusting urban planning. It can exert influence on the behavior of participants by only making a slight adjustment to the construction standards or/and the type of land development. In fact, Lanzhou municipal government is not directly involved in the development urban planning. It commissions the municipal planning bureaus, the district government as well as other relevant administrative departments, and professional planning companies to be in charge of the preparation of urban planning, but the government itself holds the rights of final approval. Therefore, it is rather easy for the government to reduce the development costs for enterprises and developers and help them raise profits. It can directly indicate its subsidiaries to specify the contents and standards of the plans or give proposed changes in the final approval procedure. Additionally, due to the lack of the involvement and review of third parties, the municipal government even can make a slight adjustment to the published construction standards and planning contents which have in the usual sense been protected by law. However, this behavior normally occurs in a private way.

Second, by virtue of a monopoly on state-owned property of urban land, Lanzhou municipal government controls the primary land market. It has authority to designate the basic price and methods of land transfer. Although Lanzhou municipal government has published the benchmark prices (roughly divided into seven levels), it can still raise/ reduce the basic standards of land leasing payment in a certain degree.

Further, the municipality is able to decide the methods of land transfer of different plots. It is also possible for it to change the planned methods. For example, lands which were supposed to be sold through a public tender, might transferred in the manner of an agreement instead. Despite private agreements or free allocations no longer being preferred by governments, they still sometimes occur. In the national and municipal documents, land transfer with an agreement has not been completely banned. Particularly in the official documents regarding urban renewal and industrial relocation and the construction of industrial parks, both central and municipal governments always give a vague description towards it. For instance, in the central government's documents[20], it is prescribed: "state-owned land (for the purpose of) is forbidden to be transferred with an agreement....some

---

20   The official documents respectively are: Notice of the Ministry of Land and Resources on Issuing the Rules on the Assignment of State-owned Land Use Right by Bidding, Auction and Quotation; and the Rules on the Assignment of State-owned Land Use Right by Agreement; data from
http://www.people.com.cn/GB/paper1787/6765/659996.html.

government-sponsored projects are allowed be transferred in this way...this includes those which containing both profit and non-profit items (greenbelt project, construction of small towns project, shantytowns, and so on), or the projects involving special high-tech and industrial uses, or those obtaining a special approval from the municipal government". This makes the land transfer with an agreement to be possible and leaves rooms for the black-box operations between local governments and other stakeholders. In 2012, Lanzhou municipal government launched the Pre-application System of Land Transfer. Its purpose is to raise funds for the primary land development in advance.

> Pre-application is much different from the public auction. It is actually the way that the municipality selects "buyers" in the office. Land applicants self-report bids. These quotations are confidential and then government decides who is enrolled — extracted from Lanzhou Morning News[21].

Another common means is that the municipal government may decide the order/time of sale of different plots by taking advantage of the land reserve system. For example, it can entrust Lanzhou Land Reserve & Investment Centre (LLRIC) and gives priority to the transaction of some designated land plots, or it can also delay the transactions. There are many reasons to illustrate these behaviors. For example, it postpones placing the plots in the market so as to assist the land transferors in profiting through land appreciation.

---

21   Quote from: http://lz.gansudaily.com.cn/system/2011/05/12/011988413.shtml.

Third, Lanzhou municipal government is responsible of collecting, managing and re-distributing local taxes. Thus, it can reduce taxes or refund part of the tax to the stakeholders, to stimulate their enthusiasm of engaging in the public-private cooperation. Means the governments used include to reduce land-related taxes and business tax for enterprises and developers or to provide enterprises with special support funds from municipal finance. Of course, this can have the opposite effect—as a kind of punishment (e.g. raising taxes and land rents). The municipal government is responsible for managing the allocated finances from the central and provincial levels. Therefore, it determines who will access to the funds.

In addition, the municipal government has authority to coordinate the works of its executive branches and governments at lower levels (i.e., municipal bureaus and district and county governments). It can command the subordinate administrative bodies to make appropriate concession to the enterprises and developers. Besides, Lanzhou municipal government is authorized officially by the central government to carry out negotiation with its higher administrative agencies (e.g. Gansu provincial government) and the administrations at the same level (e.g. the Administrative Committee of LNID)[22]. It is also possible for the municipality to require these departments and agencies to surrender part of the profits and get involved in the win-win strategy in accordance with its vision.

---

22   Administrative Committee of LNID and Lanzhou municipal government are at the same administrative level. After enterprises relocation, the relevant taxes are assigned by both administrative bodies in accordance with the ratio of 3:7.

5.3.2 A profit-distribution mechanism

Making concessions and providing incentives to enterprises

In order to urge enterprises to move out of the city center as soon as possible, the municipal government introduced a series of incentives between 2010 and 2014. First, for the enterprises that are unwilling to hand over their lands after relocation, the Lanzhou municipal government allows them to keep holding the land-use rights. But a precondition is that enterprises should pay off land conversion fees as well as other relevant taxes. Then, enterprises are available to develop the land by themselves, or transfer their land through LLRIC. The vast majority of revenue generated therefrom is owned by the enterprises.

To address the shortage of funds of relocation and workers' placement, which serves as the most reasonable excuse of enterprises, the solution of Lanzhou municipality is to provide a tax subsidy. In the compensation ordinance developed by Lanzhou municipal government, subsidies are provided in two ways. If enterprises wish to conduct the land redevelopment by themselves after the relocation, they are allowed to access to all incomes arising from the projects. The only task of them is to hand over the land appreciation tax and land conversion tax to the municipal government, which normally account for a small portion of the revenues (around 30%). In another case that enterprises transfer the land through LLRIC, LLRIC will return 75%-85% of the land transfer income to the enterprises. This proportion was just 45% before 2010.

To quell the discontents of enterprises regarding the lack of a sufficient amount of land for the establishment of new production plants and the remote locations, the municipality commissions the administrative committee of LNID to provide certain concessions on the land-transfer payment and relevant taxes. The selling price of industrial land in LNID that the committee provides to the enterprises was 225.23 yuan/m² in 2014, and the average price of commercial land was 974.99 yuan/m². This price is much cheaper than that in the main city of Lanzhou[23]. The purpose of doing so is to lure the enterprises to sell or redevelop the land in the inner city. Further, the committee offers tax relief to the relocated enterprises. For instance, the committee promises to reduce the business tax for enterprises in the early years after their relocation. This concession will terminate when the total amount of tax reduction is equal to the land purchase costs paid by the enterprises. To encourage the employees to move into LNID and simultaneously to resolve worries for the corporate executives, the committee further allocates a certain amount of lands for the construction of employees' residential and supporting facilities. These non-industrial lands are allocated to enterprises for free or at rather low prices.

Except for offering land concession and tax benefits, another means of the Lanzhou municipal government is financial incentives. In order to mobilize the enthusiasm of enterprises' executives, the municipal government promise to allocate them a sum of special

---

23   The benchmark premium of industrial land of Lanzhou main city is between 391.05 yuan/m² and 976.06 yuan/m², and the benchmark premium of commercial land is between 823.04 yuan/m² and 8,539.92 yuan/m²

construction funds, and more importantly to grant a 5-10-million-yuan reward to the executives[24].

For some enterprises with a serious lack of funds and that need loans through land mortgage, the municipality is committed to offering credit guarantees[25]. Furthermore, Lanzhou municipal government promises to assist the loss-making enterprises in upgrading the employees' living environment. The general way governments adopted is to help the enterprises to repair the dilapidated residential housing of their employees. The special measures relate to: the municipal government first designates this area as the shantytowns and lists it in the national shantytown program; it applies for the special funds from the central government; the municipal government appoints the SASAC Lanzhou Construction & Development Company (SLCDC) to undertake project financing and construction tasks; SLCDC allocated the housing to the residents and employees or resettle them in other areas. In view of the central provisions, however, the assistance of municipal is only available to the state-owned enterprises. In exchange, these enterprises asked to hand over part of their lands, and the municipality only pay the transfer fees in accordance with the standard of industrial land price.

---

24   Data is from "Measures of Industrial Enterprises Relocation and Land Reconstruction in Lanzhou" published in 2012.

25   By the end of 2013, Lanzhou municipal government has already guaranteed 500 million yuan of bank loans with urban construction lands as mortgages.

Providing incentives to developers

For the developers, one of the policy incentives that the municipal governments commonly use is to appropriately relax the restrictions on the planning indicators. In order to attract the investment of private developers in some disadvantaged locations or in the case that governments hope the developers to assume some non-profit projects during their profitable development, governments tend to relax the construction standards for them. For instance, the usual tricks of governments include raising the floor area ratio as well as building density, decreasing the greenery ratio, or even changing the originally planned uses of some plots, and so on. These acts sometimes get formal permissions by readjusting the master planning and other plans (e.g. land planning, housing construction standards, etc.). Sometimes they are privately done by the developers. In this regard, the municipal government gives its tacit agreement or sometime asks them to pay the fine symbolically. Besides, governments allow developers to construct some profitable projects in the government-sponsored programs, or relax the time limit of project completion for developers.

To address the contradictions about the pre-investment and infrastructure, the municipal government has set up specialized development companies (e.g. LIDCC and SLCDC). The companies merged the dispersed industrial plots and conduct the primary land development. This on the one hand helps the municipal government to plan large-scale projects. On the other, it contributes to reduce the upfront capital investment and manpower of developers. This cost will be passed on to the developers in the form of higher land-transfer prices. Nevertheless, it will not cause real losses to the developers, since it is not difficult to shift them on to consumers. In fact, it is a mutual-win attempt.

Furthermore, the municipal government entrusts the state-owned developmental companies (such as LIDCC and SLCDC) to offer private developers the credit guarantees of bank loans in some renewal projects. By virtue of the close relationships with state-owned banks (e.g. Lanzhou Bank, Lanzhou Branch of China Development Bank), it is not difficult for the state-owned developmental companies to gain the eligibility to provide security for bank loans. Under the government's guarantee, the private developers enable borrow the construction funds easily,. This is sometimes quite difficult if they borrow money from these banks solely relying on their own credits as well as financial strengths.

5.4 Speculative land development: from industrial space to commercial space

5.4.1 Case 1: Lanzhou LS Group Co., Ltd.

Lanzhou LS Group Co., Ltd. (LLG) is one of the biggest manufacturers in the city. It was founded by the Gansu provincial government in 1953, and was one of the 156 key national investment projects in the national "First-five Plan" (1953-1957). The enterprise was situated adjacent to the core commercial and business area in the Qilihe District, and covered an area (973,338 m²) accounting for around 6% of the total built-up area (around 15.54 km²) in the district (Figure 5.5).

Relocation debate

As early as 2007, LLG has considered the idea of expanding production land. Subject to the shortage of construction land in the surrounding area, however, it had temporarily to shelve the plan. In 2009, the Ministry of Railways and the Lanzhou municipal government jointly proposed a construction planning of Lanzhou West High-speed Railway Station. In this planning, the new railway station was chose to be close to the core commercial area of the Qilihe District—only one kilometer far away from LLG. LLG immediately realized the potential opportunity of land appreciation that the new train station will bring to it. Thus, it is one the few industrial enterprises endorsing the municipal urban renewal plan since the very beginning.

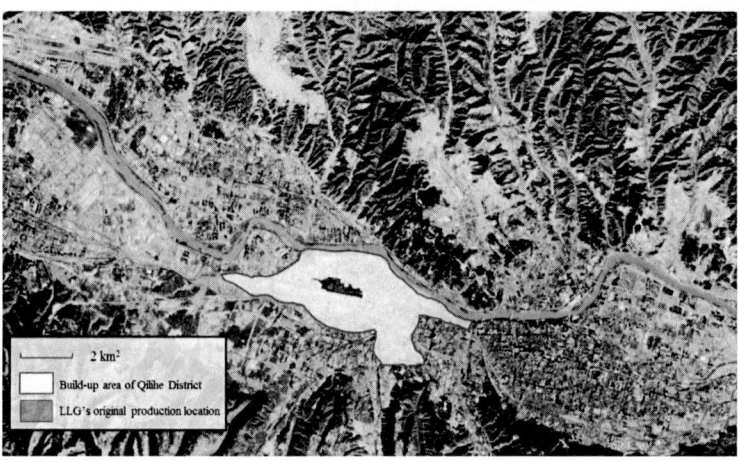

Source: drawn by the author.

Figure 5. 5 LLG's original production location and the build-up area of Qilihe District

In 2010, LLG worked actively with Lanzhou municipal government on the negotiation of relocation and siting of new production lands. In the municipal government recommendation and assistance, LLG first locked its eyes on two High-tech Development Zones respectively in the suburbs of the Qilihe District and the Anning District. However, because of sufficient construction land and infrastructure and the potential risk of a secondary relocation as these two industrial development zones are close to the inner city, LLG did not make the final decision of relocation. Another reason of the unsuccessful moving was that LLG and the Lanzhou municipal government did not reach an agreement with the distribution of land transfer incomes and the municipal compensation for the relocation. The municipal government only promised to provide new production land at very low prices, while it left the task of fund raising for relocation and resettlement to LLG itself. Besides, the municipal government required LLG to turn over all industrial lands on the original site to LLRIC for a public auction. According to the municipal expectation, the municipal government as the legal owner of the land would obtain the 65% of the land transfer income, while LLG was only allowed to access the rest 45%. Owing to the serious differences, the company withdrew its application for the relocation at the end of 2010. Although in 2011, it was listed again as one of the key relocation targets by the municipality, LLG had not made a relocation program in accordance with the government's request. The reasons it insisted not only include their previous differences, further comprising the strong opposition of workers.

We could feel the government's determination to rectify the industrial area...you see, the prices of the surrounding land and the newly developed property are rising year by year...land is state-owned, planning is developed by the municipality and the decision will be made by "the top"...relocation should be inevitable...the corporate leaders of course have recognized this, but for them, relocation may have little effects on their lives and work because they are rich...they may probably benefit from this...we have to fight for more rights as much as possible for ourselves—an interview with Hu, a retired worker of LLG in September 2013.

Since 2012, Lanzhou municipal government has issued a series of policies with incentives to urge the industrial relocation. The reward initiatives includes improving the return ratio of land transfer income to enterprises (85% of the land transfer income), allowing enterprises to conduct independent-redevelopment on the former industrial sites, providing special funds to help enterprises to upgrade technology and equipment, and carrying out financial incentives for corporate executives [26]. Punitive measures include raising the land rent on the original sites progressively, levying the costs of environmental pollution and energy consumption, and reducing the ratio of land transfer income refunded to enterprises (decreased by 15% per annum), and so on. Faced with the attractive incentives and unprecedentedly severe penalties (Table 5.1), the new site selection and the formulation of the relocation program of LLG progressed very successfully. In order to reward its positive responses on the municipal renewal project, Lanzhou municipal government granted an additional request to LLG.

---

26   Since 2011, Lanzhou municipal government allocates a sum of finance from the local financial expenditure to reward the leadership of relocated enterprises; for the enterprises which complete the relocation and the investment in new production sites punctually are able to receive a reward of 500 thousand to 1 million yuan from the municipal government.

Table 5. 1 Incentives of Lanzhou municipal government and LLG's obligations

Preferential policies from the municipal government	Obligations of LLG's commitment
LLG is awarded the land redevelopment right and the management right after the completion of projects;	LLG raises funds for the new plant construction and land redevelopment on the original site, and resolves the employee's placement after the relocation on its own;
LLG obtains 85% of the land transfer fee, and the rest 15% is allowed to pay to Lanzhou municipal government in the following years;	
Provincial and municipal governments are commitment to provide LLG with special funds (around 100 million yuan) for the project upgrading in 2010 and 2011 respectively;	LLG promised to play an exemplary role to other industrial enterprises, and to complete the relocation no later than July 2014;
Lanzhou municipal government awarded one million yuan to LLG's leadership as a relocation bonus;	LLG promised to do the redevelopment in accordance with Lanzhou City Planning (2011-2020) and Trade Circle Planning around Lanzhou West Railway Station;
The government agrees that they loan from banks with the land on the original site as collateral;	
The municipal government exempts LLG from highway fees of relocation, and promises to give the priority in providing the necessary of coal, electricity, oil, gas, transportation and other production factors to LLG, etc.	LLG turns over the land-related taxes, deed tax, business tax, etc. to the municipal government in accordance with the relevant laws and policies;…

Source: sorted out by the author based on the official documents and the related news.

According to their agreement, LLG was allowed to access all land transfer payments at once after it transferred the land on the market. Then, it was asked to refund the 15% of transfer income to the municipal government according to a fixed ratio each year.

In the previous official documents, however, all relocated enterprises are required to hand over 15% of the transfer income to the Municipal Finance Bureau after land transactions.

Under the coordination of the Lanzhou municipal government, LLG finally chose LNID as the relocation destination and signed a project agreement with the Administrative Committee of LNID in 2013. In the agreement, LLG acquired 19.77 million m² of industrial land in LNID at a price of 227.3 yuan/m². LLG did not win the land through a public auction but obtained it by a private consultation with the Committee. Although the agreement-based assignment of land-use right was repealed in Lanzhou in 2007, it is common when the Committee carries out the capital attraction in LNID. As one of the preferential measures provided by the state-level industrial zone, such an approach is tacitly permitted by central and municipal governments. The price that LLG paid for the industrial land is lower than the purchase cost of the Committee accessing agricultural land from farmers (at an average price of 330 yuan/m²).

Under the coordination of the municipal government, the Committee of LNID was committed to provide LLG with around 310,001 m² of residential land without any charges for the placement of workers[27]. The main purpose of the municipality and the committee was to assist LLG in quelling the discontent of its employees. According to the interviews with a retired employee of

---

27   Typically, LNID provides employee housing for the settled enterprises, while for a small number of large-scale enterprises, the Committee sometimes allocates a certain amount of residential land to enterprises, and the construction task is asked to be undertook by enterprises themselves.

LLG, the newly-built employee housing is sold to its current employees at a very favorable price, only one-third of the average price of urban housing in the inner city. Meanwhile, the Committee exempted LLG from the costs of preliminary land development and necessary infrastructure that shall be afforded by enterprises in accordance with the national provision. In addition, the Committee provided it with other tax cuts, such as reducing the part of the business taxes that should be turned over to LNID, and the management fees as well as import taxes of the infrastructure materials and equipment of the previous three years.

In the LLG's 2014 Annual Financial Report, the total cost of the relocation and new plant construction between 2013 and 2014 was calculated around 6.23 billion yuan, in which the investment of new plant was 6.01 billion yuan, and the relocation and construction costs were 0.22 billion yuan[28]. The vast majority of this huge expenditure, in accordance with its agreement with Lanzhou municipal government, was assumed by the enterprise itself. In recognition of LLG's active relocation, the municipal government offered 10 million yuan to enterprise's senior management as a special reward. Besides, Lanzhou municipal government applied about 100 million yuan of special funds from the central and provincial governments for LLG's equipment upgrades after its relocation. In the total amount of 6.23 billion yuan of investment, 1.26 billion yuan were collected from LLG's business income, and the remaining 4.87 million yuan of funds are from the bank loans,

---

28   Data from: the Audit Report of the Final Account of LLG's Relocation and Industrial Upgrading Program, website:
http://finance.people.com.cn/stock/n/2014/0808/c67815-25431437.html.

debt financing and issuing stocks by taking the land on the original production site as collaterals[29].

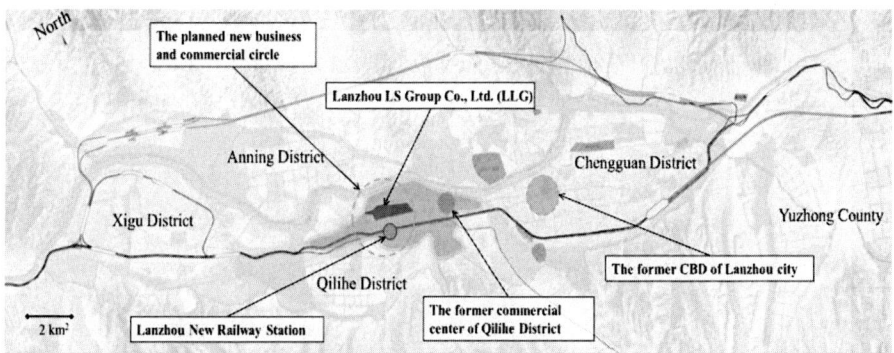

Source: drawn by the author.

Figure 5. 6 Location of Lanzhou LS Group Co., Ltd. (LLG)

Land-use debate

According to the agreement with the Lanzhou municipal government, LLG is granted the right to do independent-redevelopment on the original site after paying off 248.45 million yuan of land-use change fee as well as the relevant taxes in the following years after its relocation[30]. However in 2012, the State

---

29   LLG issued a non-public placement debt financing through the Bank of China in 2013 and successfully raised 200 million yuan. In 2014, it issued a short-term financing bonds again through ehich it raised 800 million yuan of construction funds.

30   Source: data was calculated by the author in accordance with the document of LLG Shareholders Meetings. In the document, it is pointed out that "the original industrial land of LLG was valued at 890.13 million yuan, LLG is estimated to gain 641.68 million yuan after removal of settlement costs and relevant taxes".

Council issued the Reconstruction Plan of National Old Industrial Bases (2013-2022). The Qilihe District as well as other 94 old industrial bases built in the planned economic period in the country was urged to conduct an industrial restructuring, by redeveloping the traditional industrial bases for modern business and service spaces.

Afterwards, the municipal government empowered the Qilihe district government to negotiate with LLG on the special contents of its reconstruction program to meet the requirements of the central state. However as early as 2007, the district government has in fact prepared for an industrial upgrade program which was later approved by the municipal government. In this program, the district government planned to create itself as the province's largest equipment manufacturing base, and to assist LLG as well as seven other local manufacturers in applying for the central and provincial special funds. And now, under pressure from both central and municipal governments, the district government had to give up its original plan. Owing to the inconsistent attitude of the municipality, unclear responsibilities, and the unclear benefits it could obtain, the district government's performance was not positive. The government did not participate in the formulation of specific redevelopment program, while it merely urged LLG to hand over a relocation schedule on time.

The enthusiasm of the district government was mobilized until 2013, when the Ministry of Railways and Lanzhou municipal government announced the investment project of Lanzhou West High-speed Railway Station. The project's location was chosen near the commercial center of the Qilihe district. Faced with the opportunity, the district government hoped to upgrade the center to a commercial and business zone and a new local landmark, with a joint participation of LLG and CNR Lanzhou Locomotive Ltd. (CLLL), both of which were adjacent to the new railway station and occupied a total of 1,511,538 m² of industrial land.

The final solution of their repeated negotiations is that the Qilihe district government at first designated a new district-level business center in its district land-use planning, in which the new railway station as the center. This new planned business center is at 2 km west of the original commercial center in the Qilihe district (Figure 5.6). Afterwards, the district government proposed to the municipality to upgrade the local business center to a new city-level CBD. In exchange, LLG and CLLL were respectively in charge of the commercial development of the northern and southern areas of the railway station.

Implementation of project
After LLG has paid off the land-use change fee as well as relevant taxes to Lanzhou municipal government and come to an agreement on the land-use program with the Qilihe district government, it prepared a detailed redevelopment program with the assistance of the Urban Planning, Design and Research Institute of Shanghai Tongji University in 2013 (Figure 5.7 and 5.8). In the program, there are around 0.74 km² of land being planned to be reconstructed in five years, on which LLG plans to build a commercial and residential areas called the Lanshi HOPSCA Commercial Project with a total floor area of 276.98 m² (Table 5.2 and Figure 5.9).

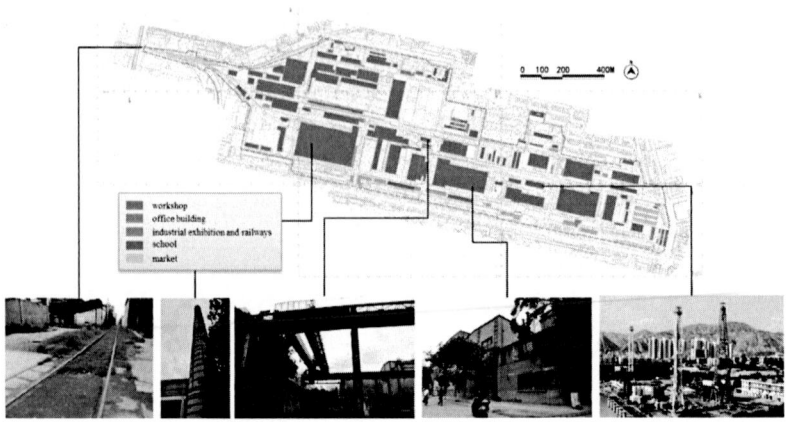

Source: photos were from field research and the map was collected from "Overall Redevelopment Planning of the LLG's Original Site".

Figure 5. 7 Land-use situation of LLG's original production site in 2013

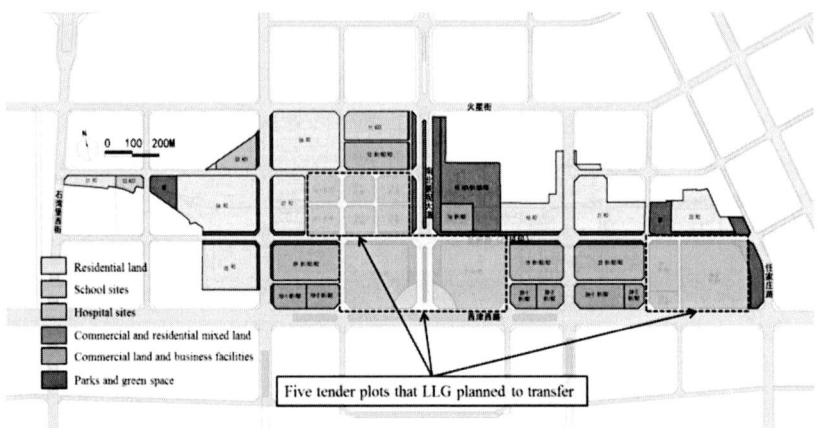

Source: drawn by the author based on the LLG's planning map.

Figure 5. 8 Redevelopment planning of the LLG's original site

Table 5. 2 Description of Lanshi HOPSCA Commercial Project

	Office Buildings	Commercial Facilities	Hotels	Residential Apartments	Culture & Creative Space	Ancillary Services
Project Description	Business center buildings; LLG's office center; creative intelligence offices; general offices	Commercial complexes; commercial services for high-speed railway station; commercial facilities facing to street	A five star hotel; business hotels; express inns.	Luxury commercial housings; general commercial housings; resettlement housing; apartments	Library and cultural center; LLG's cultural exhibition center; theater; art gallery	Primary schools; kinder-gardensh ospitals
Construction areas (km²)	0.70	0.72	0.22	1.06	0.02	0.05
Proportions	25.27%	26.15%	7.9%	38.41%	0.54%	2.21%

Source: from the "Overall Redevelopment Planning on the LLG's Original Site" published by LLG in 2013.

LLG originally planned to entrust its own real-estate subsidiary, Lanzhou Run'an Properties Ltd. (LRPL), to carry out the redevelopment projects. LRPL is formerly known as an affiliated real estate development sector which was set up by LLG at the end of 1990s. At its inception, it aimed to undertake the construction project related to the staff's welfare housing and plant facilities on the non-industrial land of LLG. Given the booming real estate market, LLG separated the affiliated department in 2005 and established it as a real-estate subsidiary with independent responsibility and autonomy. After that, the subsidiary does not only engage in the real estate development projects for its parent company LLG but also gets involved in some small-scale property projects in the city.

Source: from the Redevelopment Planning on the LLG's Original Site published by LLG in 2013.

Figure 5. 9 Effect diagram of Lanshi HOPSCA Commercial Project.

However, the real-estate subsidiary did not become one of the profitable sectors of LLG until 2013, since it only participated in a few small-scale, low-risk general commodity housing projects in the past. At the time when LLG, the municipal government and the Administrative Committee of LNID discussed on the issues relating to the relocation and land redevelopment, LLG recapitalized the subsidiary and renamed it Lanzhou LS Property Development Co., Ltd. The new subsidiary expanded its assets and business scope and was commissioned by LLG to take on all the property development projects in LNID and in the Qilihe district.

In addition, through a public auction in 2012, the subsidiary was granted the right by the Committee to engage in a development of villas and high-end commodity housing with an area of 194,000 m² of land. It obtained the piece of the land at the price of 1,315 million yuan (677.8 yuan/ m²). After the project was completed, the sales price of the commercial housing is 4,800 yuan/m². Obviously, the project helped the subsidiary and LLG get a very handsome profit. Furthermore, the subsidiary was commissioned by the Committee of LNID to be in charge of the development of LLG's employees housing (310,001 m² of residential land) in LNID.

However, faced with such a huge area of land (about 973,338 m²) on the original site in Qilihe district, the subsidiary became powerless. It could not bear all the construction tasks as its parent company LLG expected at the beginning because of the limited funds, high risk, and lack of experience. Meanwhile, LLG was also aware of the difficulties of raising the estimated 12 billion yuan of development funds in the short term[31]. With reference to new financing policies of the municipal government, as a result, LLG decided to recruit several large real-estate developers with capital

---

31   According to the enterprise estimates, the land investment needs about 7 billion yuan and construction investment demands about 5 billion yuan; and the investment is expected to generate 31.4 billion yuan of income for LLG, in which the 600,000 m² of planned commercial projects would bring about 12 billion yuan in sales, and the sale income of the office buildings and residential housing is estimated to be 19.4 billion yuan; Date from: Recommend Platform for Investment Projects in Gansu Province, http://www.investgs.org/tzxm/xdfwy/xdfwy/124.html.

capacity and rich experience to conduct a joint development on the original site. In the project description announced by LLG in March 2015, its real-estate subsidiary will only take on the development tasks of the corporate headquarters building, general residential areas and a small number of business office buildings. And five plots that were originally planned by LLG as a high-end business center and a cultural and creative industrial park are planned to be entrusted to the third-party developers (Figure 5.8). In the tender announcement, the reference price of the plot (with an area of 56,000 m²) near the main thoroughfare is 923.13 million yuan. Given the excellent location, LLG announced stringent standards. In addition, the developers are also asked to assume all development and construction costs. And after the completion of the projects, LLG requires the developers to transfer a portion of the equity of the completed commercial projects. The shares of the commercial projects that LLG occupies cannot be less than the value of the transferred land.

5.4.2 Case 2: Lanzhou Textile Co., Ltd. (LTCL)
Relocation debate and financial difficulties
Lanzhou Textile Co., Ltd. (LTCL), the former Lanzhou State-owned Textile Mill, was founded by the Lanzhou municipal government in 1958. It is located in the Xigu distrct of Lanzhou city, which has been the largest petrochemical industry base in western China since then. In 2003, the company was bankrupt; then, it was reorganized and combined with another near-bankrupt company, Lanzhou Xinglong Textile Co., Ltd. as a joint-stock company under the coordination of the municipal SASAC. However, due to the poor operation, the company was once again faced with shutdown. In 2008, LTCL had a loss of 4.22 million yuan.

Source: map was downloaded from Baidu map; photos were taken by author in 2013 and download from website: http://news.ifeng.com/gundong/detail_2013_10/15/30335847_0.shtml

Figure 5. 10 Location of LTCL's original production site and Heping Industrial Development Park

Immediately, LTCL transferred 75,333 m² of its idle industrial land to two private developers (for the commercial housing development and a small amount of supporting commercial facilities) through LLRIC respectively in 2006 and 2008 (Figure 5.10), in which a piece of land with an area of 48,866 m² helped LTCL gain an auction price of 73.5 million yuan (including taxes). However, the transfer payments are in fact far insufficient to support it in building a new plant and upgrading the technical equipment, since the vast majority of the funds were used for the placement of the laid-off workers. Due to the pressure from its employees, LTCL also asked the municipal government to put the old residential area for its workers in the agenda of municipal Shantytowns and to assume all development costs.

It is a typical industrial yard, built in the planned economy era. Most laid-off and retired workers are living in the low-cost housing of the factory, only few others bought commercial housing in other places. Although we are very desirable to improve our housing conditions, it is indeed difficult to move out of the original place in view of the high housing purchase price in other places of the cities. We understand of the financial position of the company of recent years. We must only pin our hopes on the municipal shantytowns' program. Through numerous protests of workers and the negotiation between the employee representatives, corporate leaders, and the municipal and district's governments, they have reached an agreement in 2009. The final solution is that the municipality financed the shantytowns project; and the enterprise improved the grants (10% of its original plan) for workers after the land transfer. Of course, the governments and corporate executives are the biggest beneficiaries. After the completion of the shantytowns project, they are allocated a small portion of the welfare housing for sale or for self-occupation—an interview with Jiang and Zhao two laid-off workers of LTCL in September 2013.

In the Plant Relocation and Equipment Upgrades Project signed by LTCL and Lanzhou municipal government, LTCL has pledged to invest at least 288.67 million yuan on the new site in exchange for the 80% of land transfer payments, as well as the land and tax concessions from Yuzhong County. Meanwhile, due to the worsened economic situation, it was almost impossible for LTCL to apply for a mortgage loan from the banks. As a result, the company was always in a shutdown state and had to keep on transferring the land on the original site to raise construction funds. In 2010, it once again sold a piece of land with an area of 56,667 m², and this transaction helped it raise 119 million yuan of funds (excluding 20% of the funds that had been paid to the municipal government in the form of taxes).

Due to the lack of investment funds, and thereby failing to start the construction of the new plant and equipment on time, Yuzhong County decided to withdraw the allocated land from the hands of LTCL in Heping Industrial Development Park in 2010, based on the contract they signed before. However, when LTCL stopped production, the area where it located was planned by the district government for a new commercial and residential project in the Regulatory Detailed Planning of Xigu District (2009-2022) in 2009. Many industrial enterprises (e.g. Lanzhou High Pressure Valve Co., Ltd., Lanzhou Sanmao Industrial Co., Ltd., Lanzhou Boiler Manufacturing Co., Ltd., etc.) surrounding LTCL have finished the relocation and reconstruction in line with the district plan. The slow progress of LTCL's relocation and reconstruction seriously hampered the transformation process that the Xigu district had expected.

Solutions dominated by the municipal government
Therefore, the Xigu District together with LTCL requested the Lanzhou municipal government's assistance. On the one hand, LTCL hoped that the municipal government can help it re-gain a new production site at a cheaper price than the previous one. On the other, given the municipality had risen the refund proportion

of land transfer payments from 75% to 85% LTCL wished the government would make a certain compensation for it[32].

In view of the requests from both sides, Xigu district government and LTCL, Lanzhou municipal government had to make a promise and mad a small amount compensation for it. It first returned the land-use tax of 2013 to LTCL (797,622 yuan of tax rebates) as well as other six relocated industrial enterprises that have similar experiences with LTCL. Then, the municipal government helped LTCL re-apply to Yuzhong County for around 78, 273 m² of land at almost zero cost. Third, Lanzhou municipal government entrusted the municipal SLCDC to carry out more shantytowns projects for LTCL and meanwhile to assist it in financing the relocation and reconstruction project. Under the guarantees of SLCDC, the company finally received 110 million yuan of bank loans from the Lanzhou bank and found a joint venture partner, Gansu Province Material & Industrial Group (GMIG).

In the agreement, LTCL and GMIG jointly established GMIG's Lanzhou Logistics Park Co., Ltd. (GLLP). LTCL obtained a small number of shares, in which all the land on its original site in the Xigu district and most of the land on the new site in Heping Industrial Development Park as the initial investment. In exchange, GMIG was responsible for funding. Afterwards, GLLP funded the construction of an e-commerce center with a staff residential area—

---

32   LTCL was refunded 70%-80% of land transfer payments by the municipal government after the previous three transactions. It is estimated that these transactions (283.5 million yuan) contributed approximately 70 million yuan in the form of relevant taxes to the municipal finance. Data were collected from government gazette and related news, and was calculated by the author.

providing jobs and employees' housing for both enterprises—on the LTCL's production site in Xigu District with an area of 84,042 m² (Figure 5.11). Further, it established a logistics park jointly-owned by the two companies and provided financial support to LTCL to build a textile workshop nearby in Heping Industrial Development Park.

5.4.3 Case findings and summary

By introducing two specific cases in relation to enterprise relocation and land redevelopment in Lanzhou, this section demonstrates that redevelopment process is not always plain sailing but filled with ongoing conflicts, negotiations, and cooperation. In the process, each of the major actors has its own purpose and interest demands, and equally holds different expectations for industrial areas' renewal. Yet one thing is clear: the vast majority of actors shows great interest in urban renewal projects, specifically in the potential benefits that land redevelopment will bring to them, even if many of them (enterprises) demonstrate substantial resistance toward the rejection of the municipal projects at beginning, and show an uncooperative attitude on the surface. This point is rather different from the cases of renewal of residential areas, which have been studied in detail in the literature. In the case of industrial land redevelopment, actors are bundled with each other by a variety of interest associations on the basis of land property and the potential benefits that land redevelopments are expected to bring about. In their fierce negotiations, conflicts and debated issues all focus on who is going to obtain the land-use rights after the enterprises' relocation, the destinations of relocated enterprises, the proportion of land acquisition taxes and other land-related revenues for both the original and new sites, the distributions of the project's profits after the redevelopment, and so on.

Undoubtedly, Lanzhou municipal government, which holds the unshakable authority over land ownership as well as the legitimacy of local governance concerning planning development, policy-making, projects mobilization and resource allocation, is the dominant side in this game. On the one side, the municipal government obtains certain degree of autonomy after the decentralization. If necessary, it sometimes challenges the national policies and laws that were issued by the central state. In the consensus that economic growth as a priority, the central state sometimes even acquiesced in such behaviors. For instance, the agreement-based assignment of industrial land has been prohibited by law; however, in the case of promoting enterprises located in the Fifth State-level Industrial District, such phenomena are still common.

On the other side, the municipal government hopes to preserve the economic prosperity and sustained financial growth. Therefore, it is anxious to achieve the renewal projects as soon as possible. However, its ambitious targets must simultaneously take into account the interests of other stakeholders, especially the pressures from relocated enterprises as well as their employees. For example, LLG endorsed the municipal renewal plan and showed great enthusiasm over redevelopment projects from the outset, but it was still reluctant to move until the Lanzhou municipal government promised to make a modest concession. Similarly, LPC has taken a tougher stance and refused to move until 2015, since the enterprise has been unable to reach agreement with Lanzhou municipal government on the matters about relocation costs, compensation, and the profit distribution after original lands have been transferred/redeveloped. These enterprises have once been, or are currently still contribute to the local GDP and job creation. By virtue of their strong economic ties with the local, the underlying political nexus with central or local officials, and the pressure of public opinion, enterprises have certain influences on the government's decision-making. For other enterprises those who are caught in the difficult economic and financial situations, such as LTCL, it is also possible for them to bargain with municipal government. Their negotiating power is derived from the land-use rights that they have purchased from the municipal government but are not yet expired. Although they have been unlikely to contribute to local economic growth, in order to facilitate their relocations as soon as possible, the municipal government is so generous to offer them considerable amount of production lands, financial subsidies, and poverty alleviation projects, such as Shantytown reconstruction.

However, this kind of relationships relying solely on the land property rights is extremely fragile. By grasping tightly the land ownership and unquestionable authority in urban planning, the municipal government's behavior is just like that of businessmen who claim to own the final interpretation. Thus, confronting resistance and extensive non-cooperation, Lanzhou municipal government is always able to come up with some seemingly plausible grounds to force enterprises to move out. For instance, it designated the old industrial areas as one of the transformation objects by developing urban planning and municipal strategy; or sought assistances from its superior governments to establish the

Fifth State-level Industrial District (FSID), or called on local media and the public to denounce the polluting enterprises.

Under the coercion and bribery of the municipal government, almost all stakeholders have joined the feast and succumbed to the game framework dominated by the municipal government. Stakeholders have reached a variety of short-lived partnerships on the basis of land-use rights and the generated profits that are expected from urban renewal. For example, it can be found in the alliance between the municipal government and enterprises based on their agreement over relocation costs, site selections, and placement of workers; the alliance between enterprises and property developers on the project plans and profit-sharing; the alliance between district governments and enterprises for their win-win strategies; and so on.

These alliances are unstable, within which some actors are committed to maintaining the existing relationships, while some others break the balance and seek new partnerships. In the cases of LLG and LTCL, for instance, the central and municipal states' t transformation project of old industrial areas replaced the Industrial Upgrading Program of the Qilihe district government which had been approved by the municipality a few years ago; LLG and the Qilihe district government's joint request regarding the establishment of a new city-level CBD challenged the published municipal planning; and the Yuzhong County withdrew the land that it allocated to LTCL, and LTCL and the Xigu district government jointly asked the municipal government to mediate the disputes and to provide LTCL with more subsidies. Or in the previously case of LPC, the Lanzhou municipal government blamed the central-owned enterprise—that was once seen as a loyal partner of the municipality—in the name of environmental protection while its underlying reason lies in their disagreement over relocation.

As the partnerships continued to be challenged, interests are reallocated. Renewal process is not always in compliance with the expectation of Lanzhou municipal government. Even then, unfortunately, stakeholders have inevitably succumbed to the dream that central and local governments together weaved. Deprivation and benefiting coexist. Under the mutual-win strategy designed and dominated by the municipality, obviously, central and municipal governments are the biggest beneficiaries while some others are benefit to some extent. Their common actions have finally pushed the renewal process towards a capital-driven, commercialized and speculative direction.

Chapter 6 Discussion and Conclusions

This study tries to understand the transformation of cities under the context of global and domestic politico-economic changes, and to consider how such interactions of global and indigenous ideologies and policy configuration have impacts on the urban dimension. To achieve this goal, the discussion of this study focuses on two aspects, namely the governance transition from a state-centered form to an entrepreneurial-like way; and the arising elite-led, capital-driven urban renewal projects in Lanzhou. Related scholarly research has typically blamed China's urbanization phenomenon for a shift of entrepreneurial governance, and tried to find out its institutional origins. However, many studies ignore to what extent the entrepreneurial governance in China's historical-institutional context is consistent with the one in Anglo-Saxon countries. Further, the discussions about how urban development is operationalized in an entrepreneurial-like mode and what resistance and cooperation co-existed in this process are relatively weak. Additionally, the vast majority of the case studies are concentrated in the coastal cities such as Guangzhou, Shanghai, Beijing and Shenzhen but pay less attention to the under-developed cities/regions in Western China, in which the economic output is relatively low, and the regime is affixed with a label of conservative thinking—with a weak sense of the market. In particular, there are rare studies about the governance transition in traditional industrial cities or regions which have assumed important tasks of economic production and military security in the socialist system.

In this book, I first explained why globally neoliberal discourses and general policy paradigm are selectively applied in China's system reform, and how the central state connects these concepts to its existing ideologies, politico-institutional arrangements, and the behavior criterions of actor constellations (Chapter 3.1-3.3). Then, this study examined the relation between China's institutional transition and the rise of entrepreneurial-like governance at the urban level, and the nexus between land-finance-drivers and elite-led, capital-driven urban development (Chapter 3.4). After that, this study took Lanzhou—one of the traditional industrial cities in China—as an example, to show how an urban municipality with the assistance of central and provincial governments as well as subsidiary departments, initiated the renewal projects of traditional industrial areas, coordinated the conflict of interests among stakeholders, and ultimately achieved the goal of urban renewal (Chapter 4 and 5).

6.1 Empirical findings and discussion

China's neoliberal steering with a "selective" logic

In Chapter 3, I adopted a perspective of institutional analysis approach to examine China's institutional transition. It is argued that neoliberalism and China's indigenous ideologies have been alternately dominant in the arena of public opinion in regard to its political reform. This phenomenon is accompanied by the repeated negotiations among various interest groups, including the reformers and conservatives among political elites, the social organizations, groups and citizens, as well as the newly emerging economic elites along with the rise of domestic capital market.

I prefer to deem China's institutional transition a "neoliberal steering with a selective logic." This selectivity has a double meaning. At first, the open and market-oriented path is a strategic choice of the political elites, when the whole country had suffered from severe economic recession, political instability and social unrest throughout the 1960s and 1970s. More precisely, it is the result of the conservatives temporarily compromising with the reformers as well as the growing capital forces. However, this strategic choice is obviously limited in degree. The decisions that political elites made are more or less limited by their ideologies (i.e. communism) as well as the specific historical and cultural background. Then, China's further neoliberal steering should be attributed to a large extent to the penetration of global neoliberalism into China's indigenous ideologies as well as its socio-economic regimes. It is enhanced by global powerful discourses as well as hegemonic international norms and standards, under the condition of increasingly frequent economic and trade exchanges between global and local.

In this chapter, I also examined how China connects neoliberal concepts to its existing politico-institutional arrangements, as well as its purpose in doing so. I argued that although China has carried out a series of attempts of neoliberal practices, its reform initiatives are far different from those being implemented in a violent or radical way in the 1990s' Eastern Europe as well as Latin America. On the one hand, the political elites announced that only if China takes a "gentle, peaceful and gradual" approach to the reform, the state will it be possible to avoid economic collapse as well as social chaos. On the other, central government repeatedly corrected its definition of the market in the previous CPC's National Congresses. It introduced the market mechanism in some carefully selected areas (urban housing and land systems, the financial system, etc.)

progressively. The aim is both to ensure the legitimacy of the ruling party (i.e. to maintain its authority in economy and society), and to seek assistance from capital market as well as civil society to promote economic prosperity.

A corollary is that administrative power is bound to be weakened, but it has not vanished. Instead, direct government regulation is replaced by the alternatives, exerting influence in a more indirect and subtle way. The most typical approach is that governments, on the one hand, place urban land in the market, while on the other firmly grasp land ownership in hand. It gives governments the justification for interventions, in the name of maintaining a stable and equitable market. The promotion of housing privatization and the regulation of the housing market follow the same logic. Another neoliberal approach with Chinese characteristics is that the central government is committed to maximize the enthusiasm of local governments through decentralization; while at the same time it constraints behavior of local officials by enacting a strict cadre appraisal system. As such, the central government allows local authorities enjoy a certain degree of autonomy while exerting subtle influence on the behavior of local officials via Cadre Appraisal System.

The rise of entrepreneurial-like governance
In this study, the selective adoption of the ideas of neoliberalism is considered as the fundamental genesis of the universal "contradictory duality" existing in many spheres of China's politico-institutional system. And all this is attributed to the root cause of the changeable and contradictory urban policies, multiplicity of governance regimes, and the consequently unique

urbanization processes filled with various conflicts and struggles over interests and rights.

By analyzing China's urban land and housing system reform, fiscal system reform, as well as the deliberate design of central government's deliberate design of a cadre appraisal system, this study comes to the following conclusions. The restructured institutional system acting as a set of new rules of the game has reconfigured public resources by re-defining property rights. Meanwhile, it has reshaped interest relations by defining the way of accessing resources, as well as power structures by re-scaling the central-local relation. Precisely because of the re-combined rules, which function as Foucault deemed the "conduct of conduct" of actors, the role and obligations of local governments have been repositioned, and the ways of their power operation and resource allocation are also changed. Thus, a certain degree of decentralization of powers and responsibilities contributes to the steering of local governance towards entrepreneurialism, whilst the imbalance between financial revenues and local responsibility makes this a governance model characterized by over-reliance on land finance and a pursuit of maximization of land profits. Accordingly, to maximize land value through the restless land (re)development and infrastructure investment has naturally turned into an effective and prevalent strategy among the vast majority of Chinese cities and towns.

Even if this land-targeted governance has obvious entrepreneurial features, in this study, I call this "entrepreneurial-like governance." By studying the case of Lanzhou (Chapter 4), I found that although local authorities have been given greater autonomy than ever before, the administrative implications of the central on municipal authorities has not completely disappeared. Apart from affecting the behavior of local governments by re-adjusting the rules of the game (i.e. politico-institutional arrangements) as described above, there exists a kind of weak yet still effective administrative interventions from the central state. These subtle interventions are primarily embodied in two aspects. They are respectively related to the indirect effects of national economic objectives as well as urban development strategies on the target-setting and policy-making of local authorities; and the macro policy directives on local development, such as national strategies as well as relevant policies, regarding Shantytowns, old industrial areas' renewal, and the experimental market-oriented reform around urban management and construction (Chapter 4.5).

An alliance-based development

At last, this study adheres to the necessity of understanding the multiplicities of local governance in the context of China's transforming metropolises. That is to say, reflections on urban governance should go beyond the analytical framework that governance regimes are considered as singular and mutually exclusive. This disrupts the notion that China's local authorities are the singular protagonist within a static and suppressed social structure and that they play an undisputed hegemonic role in the rapid urban transformation.

In the case study of Lanzhou's traditional industrial area (Chapter 5), my study indicates how urban transformation is affected by interactions among multiple stakeholders—i.e. governments at all levels, relocated enterprises, and property developers. Their relationships cannot be described as purely antagonistic—i.e. domination and resistance, or leading and being dominated. On the contrary, since the redistribution of property rights, obligations, interests, and so on after the reform, stakeholders have been connected to each other by intricate chains of interests as well as various associated liabilities. This means the balance of power structures as well as the various interest relations based on the existing governance framework have been partly broken, while alternatively a set of new relationships has been established step by step. Thus, in this mixed-configuration of relationships, apart from constant conflict and resistance, instead, there are a wide range of competition, negotiation, compromise and cooperation.

Chapter 5 presents a scene of conflicts and struggles among stakeholders, whose legitimacy of participating in the renewal projects are respectively specified by the old and emerging interest frameworks at the local level. Through two specific cases, it reveals an "interest-distribution" mechanism, which is dominated by Lanzhou municipal government for the purpose of resolving the conflict, and pursuing a "win-win", profit-driven target. It argues that through this mechanism, all stakeholders have eventually voluntarily or involuntarily gotten involved in the speculative redevelopment processes, and reached various alliances for common interests. However, due to the inherent instability of this mechanism, characterized by unilaterally-dominant, semi-compulsory and a lack of multiple-cooperative-targets, the partnerships in line with government's conception appear to be temporary and rather fragile. Hence, accompanied by a series of explicit and implicit competitions and resistance as well as repeated negotiations within these alliances, the established interest relationships are constantly reshaped, which thereby affects the methods and consequences of the production of urban space.

6.2 Inspirations

In the conclusions, therefore, it can be noticed that, either about the mobilization of urban renewal, or regarding the ways of organization and development, there are still somewhat "top-down" features. Just this sort of "downward penetration" of administrative authority becomes more weak and subtle. In the municipal government-led framework (including objectives and rules of land development, mutually beneficial mechanism, cooperative approaches and forms of partnerships, etc.) relating to urban renewal, apparently, the "free" choices of stakeholders are interfered, and to some extent comply with the grand blueprint of national economies, and the concept of national and local governments on social harmony and urban development.

When the central joined with local governments, and "selectively" leveraged the market as well as capital forces, arguably, the vast majority of stakeholders have involuntarily fallen into the speculative urbanization processes, succumbing to the state's ambitious visions and the logic of capital accumulation. Consequently, the power of the individual appears to be negligible, and actors are bound to yield to the powerful union of state and market. Thus, as we have seen, "tabula rasa" approaches have been adopted in renewal of old urban fabrics at an incredible rate and in a broad geographical scope within the country. Through such extreme means of quick success, China's urban transformation is characterized by several distinctive features, namely benefit-oriented development, restless land speculation, and an excessive commercialization of the space. During the process of Lanzhou industrial area renewal, in addition to the seemingly attractive "win-win" strategy and generous benefit-sharing, there are also the surging undercurrent of relentless deprivations and violations. Hence, the issue of whether such urban development path is sustainable is urgently debatable.

That being the case, we cannot help but reflect on whether there is an alternative possibility regarding urban governance to eliminate, or to gradually resolve such a series of contradictions appearing in current land (re)development and urban transformation, which should be fundamentally attributed to the "paradoxical coexistence" of China's hybrid institutional framework, or probably at a deeper level, to the ongoing ideological confrontations as well as integrations over the last 40 years. The answer is that, it is not impossible. It is nothing less than an optional choice that to eliminate the internal defects and universal contradiction through a progressive adjustment of current politico-institutional arrangements. Nevertheless, it appears to be unrealistic to completely sweep away the interferences of the conflicting ideologies—in fact, no ideology exists in isolation, or to come up with a set of near-perfect institutional arrangements and policy configuration within a short time. No matter ideological conflicts or institutional evolution, they are always accompanied by the endless struggles and negotiations between different interest groups. It is truly hard to say which side will be conquered by others in the true sense, but is only suppressed in a certain time. That is to say, the "paradoxical coexistence" is endogenous and simultaneously eternal. Hence, rather than expecting immediate changes at the ideological and institutional dimension, a sort of "bottom-up" interference seems to be more realistic and achievable, and is more likely to yield results in the short run.

Accompanied by the "top-down" process that "information" flows from the abstract ideologies to the concrete institutional arrangements and then affects the governance pattern as well as policy-making at the urban level, which thereby (re)shaping the interactions of stakeholders, there exists simultaneously a kind of "upward" flow of "information" (see Figure 1.1 in Chapter 1). Such information feedbacks are passed up through various sorts of interactions and relationships, sometimes in the form of competitions and struggles between the groups seizing the initiative and the disadvantaged groups, while sometimes through their negotiations and cooperation. It thereby orients the transformation of urban governance and decision-making (involving the role and responsibilities of actors, the ways and basis of stakeholders associated with each other, and so on), and may also indirectly influence the evolution of a national institutional system. Although the latter effects are seemingly slight, this does not mean it is not possible.

In the case of Lanzhou, it is noticed that although the central interventions at local and its authority in urban governance still exist in some degree, their influence has been greatly weakened, and the means of intervention becomes increasingly indirect and subtle. Precisely as the looseness of the "top-down" authority, a kind of "bottom-up" and spontaneous forces has emerged, with the fast expansion of market forces as well as place-based forces. Their growth keeps on challenging the "top-down" imposed influences, and further compels a "self-adjustment" of the old hierarchy, such as the resistance from local government to the central state, and from wide social forces (mainly referring to the strength from economic sectors) to local authorities.

In the context of this study, the fading of political ties between enterprises and local governments after the SOEs reform and the reform of property rights system (e.g. land system reform) have cultivated a number of local economic entities with independent decision-making powers, and whose private property is currently under the protection of law. Faced with the municipal strategies as well as planning that undermine the rights and interests of relocated enterprises, they have strength to negotiate with local authorities, and even have reason to challenge the legitimacy of the strategies, at least nominally. Thus, to some degree, they are able to exert influence on the plans and contents regarding municipality-led projects, bargain with governments at all levels, and fight for more rights and interests which are beyond the original plans of municipality.

For instance, the Lanzhou municipal government made a concession that raising the proportion of the return of land transfer for enterprises; it was committed to undertaking the construction tasks of shanty towns for some enterprises who were trapped in troubles of employee relocation and housing subsidies; governments at central and municipal levels promised to offer specific funding as the subsidies of technological transformation for polluting enterprises; and so on. In their negotiation processes, the interests of third parties are sometimes taken into account and are utilized as a bargaining chip by some parties. For example, the municipal government struck back enterprises that refused to move, in the name of environmental protection and social responsibility; or some enterprises asked the municipal government to undertake part of the resettlement compensation—normally in the form of raising the return of land transfer payment to enterprises, or to provide affordable housing for their employees. Even if the motivation is questionable, the community and the surrounding residents are indeed able to benefit from it.

Based on these findings, this "bottom-up" alternative can be envisaged to create possibility and provide legal protection for a more extensive involvement of social groups that represent those with different objectives and demands (e.g. enterprise employees and nearby residents, and independent planning companies). That is to say, a new mechanism that does not exclude any stakeholders in any forms is urgently advocated. A broad multilateral participation, on the one side, is conducive to getting rid of the unilateral monopoly through numerous benefit struggles and extensive negotiations. On the other, the multiple objectives of participants will give rise to the diversity of urban development. Besides, ensuring the legality of participation is also necessary, in which clarifying and protecting property rights would be an effective approach. It to some extent ensures the coexistence of different voices on policy-making and urban development at the legal level, and it enables stakeholders to be active actors in this process by linking their tangible benefits to urban transformation.

Additionally, in view of the uneven strength in negotiations, wherein both discursive and administrative powers of the state and capital are far ahead of other social forces (e.g. relocated enterprises, employees, nearby residents, and even media), the way of seeking a universal "place-based alliance" is practical to protect the vulnerable groups. For instance, the alliances being formed between corporate leadership and employees on the basis of welfare security agreements, and among internal employees of enterprise for common interests, to fight the compulsory demolition and the unjust relocation compensation. Another example is the alliance forged by the municipal government and the public to fight back the tough stance of the central state-owned enterprises as well as their political backgrounds.

6.3 Theoretical and practical implications
Theoretical implication
This study contributes directly to the sub-fields of urban and development geography. In the course of study, I attempted to conceive of an analysis framework that integrates elements from the neo-Marist political-economic approach and Foucauldian discourse analysis. This hybrid approach has been used by a few scholars in the past few years but still leaves sufficient room for further studies (Mayer and Künkel, 2012; Uitermark, 2005; MacLeod and Jones, 1999). In particular, it lacks of concerns over those with a more sophisticated, conflicting cultural and political context, for instance the cases involving some hybrid regimes (authoritarian, democratic and communist regimes) in the Global South. The purpose of taking this hybrid analytical framework is to overcome the shortcomings of each perspective (Mayer and Künkel, 2012, p4), and to highlight their strengths as a complement to the other side to present a more systematic interpretation of the complex phenomenon of urban neoliberalization. That is to say, the primary tasks are to pay equal attention to the genesis and evolution of institutions that orient urban policies and collective actions of the actor constellation towards a localized neoliberalization, on one side; and to the "microscopic mechanism" of governance hidden behind this process, which plays an essential

role in the form of specific techniques and knowledge to facilitate the trajectory of evolution, on the other side.

All in all, the political economy approach in this study is used to answer why the identity of different subjects have been (re)shaped (i.e. the central state, local governments and the market), and what changes have happened to their relationships in the process. It paid particular attention to the dominant side in the restructured power structure—i.e. who own the power and how they are shaped in institutional transition. While Foucauldian governmentality approach is employed to parse and reveal the (re)formation processes of power structure and the microscopic procedure of power operating—i.e. how different subjects exercise their power, and how the exerting power and being dominated by the power constitute the links between each subject.

Empirical implications

By drawing on the Evolutionary Game Theory as a standing point, Chapter 3 answered the selective logic of China's neoliberal steering is an outcome of the repeated negotiation among different interest groups whose political stands are interfered by the penetration of different ideologies. Along with the constant shift in the power of interest groups, the idea of neoliberalism and authoritarianism dominate alternately in China's political arena. This is the fundamental genesis of China's institutional transformation that is advancing haltingly towards an orientation of marketization, liberalization and deregulation. Moreover, the concept of path dependence was reference to illustrate the state's selective strategy towards the neoliberal paradigm—i.e. why the central state still maintains a dominant position relative to the market and local governments. As such, this study presents a hypothesis that the specific institutional arrangements are constructed on the basis of a set of power structure in a given cultural and political context, in which the dominant party of various interest groups has a crucial impact on the trajectory of institutional evolution. Additionally, from a political-economic perspective, the cause of local state transition concerning their roles and duties has been discussed in detail in Chapter 3. It is argued that local state transition is closely accompanied by the changing rule of the game (i.e. what resources are available to them, and the way of accessing to local resources) and the varying power and interest relations (i.e. the relationships between governments and the market and between the central state and local government) associated with a macro-institutional transition.

In the case study, I paid an equal attention to the concept of Foucault's the "conduct of conduct". Foucauldian analysis approach is exploited to explain an important issue—i.e. how power is exercised by the subject. Specifically, it has been verified in two aspects. First, it is adopted to answer how state hegemony functions in affecting urban governance and local decision-making. In Chapter 3, it is demonstrated how the central authority makes adjustments to the established institutional arrangements deliberately and progressively, whose purpose is rather clear, on the one hand, to maintain the existing power structure (i.e. to protect the central hegemony), while on the other to moderately delegate its power (i.e. to establish the market and the self-responsible local governments being subjected to the administrative controls) in response to the varying socio-economic environments. Then, this approach is used to examine the way (or techniques) of local authorities in the mobilization, coordination and organization of mega-renewal projects and commercialized land development (Chapter 4 and 5). In Chapters 4 and 5, in addition to revealing the strategies and techniques that local authorities innovated and exploited to driven urban renewal, the issue about how stakeholders are eventually voluntarily or involuntarily involved in the capitalized urbanization with the municipality's formal and informal mobilization, and make contribution to the neoliberalization process at the local level, was outlined.

Through the above theoretic and methodological assemblage, one of the core issues of the study has been illustrated—that is, how are globally circulating discourses and policy paradigms selectively chosen, transformed and presented locally? About another important issue—i.e. to what extent the neoliberal discourses and policies have been applied locally, it has been answered through a careful investigation of the ubiquitous administrative implications. This study endorses the notion that "oppression and resistance", but does not agree with the view of a relatively eternal and static state of binary oppositions-- the oppressed is always in compliance with hegemonic rule. It is more inclined to the conclusion that: in addition to oppression and resistance, there are also constant competition, negotiation, compromise and cooperation; and the status of hegemonic power is simultaneously being challenged.

6.4 Shortcomings of the research and prospects

Because of the limitations about time and survey methods, this study has some shortcomings. First, this study is conducted around the discussion of the relationship between the state and the market, while it involves fewer discussions about social aspects. In the empirical study, the investigation and study of enterprise workers, nearby residents, and other vulnerable groups are inadequate. It is not without regret that the study failed to consider them as one of the main stakeholders and to integrate their demands and negotiations in the analysis framework. Second, in the inspiration section of Chapter 6.2, a kind of possible alternative has been mentioned, namely the "bottom-up", spontaneous forces to challenge the "top-down" hegemonic penetration, which is thereby expected to put pressure on the existing power framework. However, this study failed to provide more detail case studies to verify the information feedback (see Figure 1.1 in Chapter 1).

These inadequacies of the study have also shed light on future research directions. Extending from the research contained within this book, it can be expected to create further discussion on the social and spatial effects that the elite-led, capital-driven urban renewal programs and speculative land development have brought about, particularly, by focusing on people's experiences of urban transformation in the shifting practices of social reproduction under state and market transition.

Bibliography

Akçalı, E., & Korkut, U. (2015). Urban transformation in Istanbul and Budapest: neoliberal governmentality in the EU's semi-periphery and its limits. *Political Geography*, *46*, 76–88. http://doi.org/10.1016/j.polgeo.2014.12.004

Ambrosino, A. (2013). Institutions as game theory outcomes: toward a cognitive-experimental inquiry. *International Journal of Management Economics and Social Sciences*, *2*(2), 129–150. Retrieved from http://www.ijmess.com

Anderberg, S. (2004). Systems Analysis in Geography. In Olsson, M.O., & Sjöstedt, G., *Systems Approaches and Their Application*. Netherlands: Springer, 79-93.

Anguelov, D. (2015). *Financialization of entrepreneurial urbanism: neoliberal governance, governmentality, and path-dependent restructuring in the European Union's cohesion policy.* University of California, Los Angeles. Retrieved from http://escholarship.org/uc/item/9j22s0tw

Aoki, M. (1998). The subjective game form and institutional evolution as punctuated equilibrium. In *Second World Congress of the International Society for New Institutional Economics*. Paris.

Aoki, M. (2001). *Towards a comparative institutional analysis*. MIT Press, Cambridge, MA

Arendt, H. (1968). Introduction: Walter Benjamin: 1892-1940. *In Illuminations*. New York: Schocken Books.

Arnould, E.J., & Thompson, C.J. (2002). Consumer Culture Theory (CCT): Twenty Years of Research. *Chicago Journal*, *31*(4), 868-882.

Beatty, R. (2014). Neoliberal urbanism: socio-spatial fragmentation & exclusion. *New Visions for Public Affairs*, *6*.

Beck, U. (2000). *What is globalization?*. Cambridge: Polity Press.

Bell, S. (2002). Institutionalism-old and new. In D. Woodward (Ed.), *Government, Politics, Power and Policy in Australia* (7th ed.). Melbourne: Longman.

Belle, I. (2015). *From economic zone to eco-city? : Urban governance and urban development trends in Tianjin's coastal area*, (Eds.) by Gebhardt, H., Stuttgart: Borntraeger.

Bentley, G., Shutt, J., & Pugalis, L. (n.d.). LEPs: Entrepreneurial governance, soft state spaces and innovative practices.

Bian, Y., & Logan, J. R. (1996). Market transition and the persistence of power: the changing stratification system in urban China. *American Sociological Review*, 61 (5): 739- 758.

Birkmann, J., Garschagen, M., Kraas, F., & Quang, N. (2010). Adaptive urban governance: new challenges for the second generation of urban adaptation strategies to climate change. *Sustainability Science*, 5(2), 185–206.

Boettke, P. (1990). The theory of spontaneous order and cultural evolution in the social theory of FA Hayek. *Cultural Dynamics*, 3(1), 61–83. http://doi.org/10.1177/092137409000300105

Brenner, N. (2000). The urban question as a scale question : reflections on Henri Lefebvre, urban theory and the politics of scale. *International Journal of Urban and Regional Research*, 24(2), 361–378. http://doi.org/10.1111/1468-2427.00234

Brenner, N. (2004). Urban governance and the production of new state spaces in western Europe, 1960–2000. *Review of International Political Economy*, 11(3), 447–488. http://doi.org/10.1080/0969229042000282864

Brenner, N. (2009). What is critical urban theory?. City, 13(2), 198-207.

Brenner, N., Marcuse, P., & Mayer, M. (2009). Cities for people, not for profit, *City*, 13(2), 176-184.

Brenner, N., Madden, D., & Wachsmuth. D. (2011). Assemblage urbanism and the challenges of critical urban theory. *City*, 15(2), 225-240.

Brenner, N., Peck, J., & Theodore, N. (2010). After neoliberalization? *Globalizations*, 7(3), 327–345. http://doi.org/10.1080/14747731003669669

Brenner, N., & Theodore, N. (2002a). Preface: from the "new localism" to the spaces of neoliberalism. *Antipode, 34*(3), 341–347. http://doi.org/10.1111/1467-8330.00245

Brenner, N., & Theodore, N. (2002b). Cities and the geographies of "actually existing neoliberalism." *Antipode, 34*(3), 349–379. http://doi.org/10.1111/1467-8330.00246

Butler, T., & Robson, G. (2001). Social capital, gentrification and neighbourhood change in London: a comparison of three south London neighbourhoods. *Urban Studies, 38*(12), 2145–2162. http://doi.org/10.1080/00420980120087090

Campbell, J. (2007). The rise and transformation of institutional analysis. *International Center for Business and Politics, 10*. Retrieved from http://laisumedu.org/DESIN_Ibarra/nuevoinst2007/borradores/Campbell.pdf

Castells, M. (1979). The urban question: a marist approach. London: Edward Arnold (Pblishers) Ltd., 1979.

Castree, N. (2006). From neoliberalism to neoliberalisation: consolations, confusions, and necessary illusions. *Environment and Planning A, 38*(1), 1–6. http://doi.org/10.1068/a38147

Castree, N. (2008). Neoliberalising nature: the logics of deregulation and reregulation. *Environment and planning A*, 40(1), 131-152.

Certeau, M. de. (1984). *The practice of everyday life. University of California Press*. Berkeley: University of California Press.

Chang, T. (1999). Local uniqueness in the global village: Heritage tourism in Singapore. *The Prefessional Geographer*, 51(1), 91-103.

Chang, T. (2014). "New uses need old buildings": gentrification aesthetics and the arts in Singapore. *Urban Studies, 53*(3), 1–16. http://doi.org/10.1177/0042098014527482

Chen, A. (1998). Inertia in reforming China's state-Owned enterprises: The case of Chongqing. *World Development*, 26(3): 479-495.

Choi, N. (2016). Metro Manila through the gentrification lens: disparities in urban planning and displacement risks. *Urban Studies, 53*(3), 577–592. http://doi.org/10.1177/0042098014543032

Chu, Y. (2002). Re-engineering the developmental state in an age of globalization: Taiwan in defiance of neo-liberalism. *China Review*, *2*(1), 29–59.

Chuang, J. (2015). Urbanization through dispossession: survival and stratification in China's new townships. *Journal of Peasant Studies*, *42*(2), 275–294. http://doi.org/10.1080/03066150.2014.990446

Clarke, S. E. (2007). Embedding governance arrangements. In *Conference on A Global Looks at Urban and Regional Governance, Halle Program on Governanc* (pp. 1–36). Atlanta: Emory University.

Clarke, S. & Gaile, G. (1989). Moving toward entrepreneurial economic development politics: opportunities and barriers. *Policy Studies Journal*, 17 (spring): 574-598.

Commons, J. R. (1934). *Institutional economics: Its Place in Political Economy*. New York: Macmillan.

Cox, K.R., & Mair, A. (1988). Locality and community in the politics of local economic development. *Annals of the Association of American Geographers*, 78 (2), 307-325.

Difaetano, A., & Strom, E. (2003). Comparative urban governance an integrated approach. *Urban Affairs Review*, *38*(3), 356–395. http://doi.org/10.1177/1078087402238806

Ding, C. (2003). Land policy reform in China: assessment and prospects. *Land Use Policy*, 20 (2), 109-120.

Douglas, M. (1986). How institutions think. New York, USA: Syracuse University Press. Retrieved from http://books.google.co.uk/books?id=oFnWbTqgNPYC\nhttp://books.google.com/books?hl=en&lr=&id=oFnWbTqgNPYC&oi=fnd&pg=PR5&dq=Institutions,+institutional+change,+and+economic+performance&ots=sXkvO8KjR3&sig=fW1ICy5AIuoNgq2hYjzHlW97_

Driessen, P. P. J., Dieperink, C., van Laerhoven, F., Runhaar, H. A. C., & Vermeulen, W. J. V. (2012). Towards a conceptual framework for the study of shifts in modes of environmental governance - experiences from the Netherlands. *Environmental Policy and Governance*, *22*(3), 143–160. http://doi.org/10.1002/eet.1580

Duckett, J. 2001. Bureaucrats in business, Chinese-style: The lessons of market reform and state entrepreneurialism in the People's Republic of China. *World Development*, 29 (1), 23–37.

Duckett, J. (2004). State collectivism and worker privilege: a study of the urban health insurance reform. *China Quarterly*, 177, 155-173.

Eick, V. (2006). Preventive urban discipline: rent-a-cops and neoliberal globalization in Germany. *Social Justice*, 33(3), 66-84.

Edin, M. (2003).State Capacity and Local Agent Control in China: CCP Cadre Management from a Township Perspective. *The China Quarterly*, 2003(173), 35-52.

Elden, S. (1998). Henri Lefebvre and the production of space. In *PostModerne Diskurse zwischen Sprache und Macht* (pp. 1–12). Retrieved from http://www.gradnet.de/papers/pomo98.papers/stelden98.htm

Elden, S. (2001). Politics, philosophy, geography: Henri lefebvre in recent Anglo-American scholarship. *Antipode*, *33*(5), 809–825. http://doi.org/10.1111/1467-8330.00218

Enright, T. (2012). *Building a grand Paris: French neoliberalism and the politics of urban spatial production*. University of California – Santa Cruz.

Etzold, B., Jülich, S., Keck, M., Sakdapolrak, P., Schmitt, T., & Zimmer, A. (2012). Doing institutions. A dialectic reading of institutions and social practices and its relevance for development geography. *Erdkunde*, *66*(3), 185–195. http://doi.org/10.3112/erdkunde.2012.03.01

Feldman, E. M. (2001). *The emergence of a new governance system in Argentina: institutional change, politics and economics. The University of North Carolina at Chapel Hill*. The University of North Carolina at Chapel Hill. Retrieved from http://search.proquest.com/docview/60673496/abstract/embedded/W WP5895J7MDINB4D?source=fedsrch

Florida, R., Mellandder, C., & Qian, H. (2012). China's development disconnect. *Environment andPlanning A,* 44,628–648.

Fonza, A. H. (2010). *Troubling city planning discourses: a womanist analysis of urban renewal and social planning in Springfield, Massachusetts, 1960-1980.* University of Massachusetts-Amherst.

Foucault, M. (1977). The eye of power. In Gordon, C. (Eds.), *Power/Knowledge: Selected Interviews and Other Writings,* 1972-1977. New York: Pantheon, 1980, 146-165.

Foucault, M. (1988a). Technologies of the Self. In Martin, L., Gutman, H. & Hutton, P. (Eds.) *Technologies of the Self: A seminar with Michel Foucaul.* Amherst: The University of Massachusetts Press, 16-49.

Foucault, M. (1988b). Truth, Power, Self: An interview with Michel Foucault. Rux Martin. In Martin, L., Gutman, H. & Hutton, P. (Eds.) *Technologies of the Self: A seminar with Michel Foucault.* Amherst : The University of Massachusetts Press, 9-15

Foucault, M. (1984). Of other spaces: utopias and heterotopias. *Architecture /Mouvement/ Continuité,* 1–9.

Foucault, M. (2008). *Birth of Biopolitics.* New York: Palgrave Macmillan.

Frederickson, H. G., and Smith, K. B. (2003). *The Public Administration Theory Primer.* Boulder, CO: Westview Press.

Fu, Q. (2015). When fiscal recentralisation meets urban reforms: Prefectural land finance and its association with access to housing in urban China. *Urban Studies,* 52 (10), 1791-1809.

Gaubatz, P. (2002). Looking west towards Mecca: Muslim enclaves in Chinese frontier cities. *Built Environment, 28*(3), 231–248.

Geciené, I. (2002). The notion of power in the theories of Bourdieu, Foucault and Baudrillard. *Sociologija,* 116–125.

Goh, L., & Williams, P. (2011). Neoliberalism and the role of the state in surplus public land management: protecting Sydney Harbour's open space legacy. In *State of Australian Cities National Conference.* Retrieved from http://apo.org.au/node/60006

Goldman, M. (2011). Speculative urbanism and the making of the next world city. *International Journal of Urban and Regional Research, 35*(3), 555–581. http://doi.org/10.1111/j.1468-2427.2010.01001.x

Golubchikov, O., Badyina, A., & Makhrova, A. (2014). The hybrid spatialities of transition: Capitalism, legacy and uneven urban

economic restructuring. *Urban Studies, 51*(4), 617–633.
http://doi.org/10.1177/0042098013493022

Golubchikov, O. (2010). World-city-entrepreneurialism: globalist imaginaries, neoliberal geographies, and the production of new St Petersburg. *Environment and Planning A, 42*(3), 626–643. http://doi.org/10.1068/a39367

Goonewardena, K., Kipfer, S., Milgrom, R., & Schmid, C. (2008). *Space, difference, everyday life. Routledge.* New York and London: Routledge. http://doi.org/10.1111/j.1745-7939.2009.01167_6.x

Gough, V. J., Eisenschitz, A., & McCulloch, A. (2006). *Spaces of Social Exclusion.* Psychology Press.

Gramsci, A. (1971). *Selections from the Prison Notebooks of Antonio Gramsci,* New York, International Publishers.

Greif, A. (1993). Contract enforceability and economic institutions in early trade: the Maghribi traders' coalition. *Am Econ Rev,* 83(3),525–548.

Greif, A. (2006). *Institutions and the path to the modern economy: lessons from medieval trade. Cambridge University Press.* Cambridge, UK: Cambridge University Press. http://doi.org/10.3917/rpec.102.0121

Greif, A., & Kingston, C. (2011). Institutions: rules or equilibria? In *Political Economy of Institutions, Democracy and Voting* (pp. 13–43). Berlin, Heidelberg: Springer-Verlag. http://doi.org/10.1007/978-3-642-19519-8_2

Grief, A., & Laitin, D. D. (2004). A theory of endogenous instituitonal change. *American Political Science Review, 98*(4), 633–652.

Gross, J. S. (1999). *Urban political participation in London: the impact of institutional transformation.* The City University of New York.

Gürpınar, Ö. (2014). On the trace of the neoliberal governmentality: a case study on the massive urban projects in Istanbul. In *Berlin and Istanbul Lecture Series.* Istanbul.

Hackworth, J. (2007). *The neoliberal city: Governance, ideology, and development in American urbanism.* Ithaca & London: Cornell University Press.

Haider, D. (1992). Place war: new realisties of the 1990s. *Economic Development Quarterly,* 6(2), 127-134.

Hall, P. A., & Taylor, R. C. R. (1996). Political science and the three new institutionalisms. *Political Studies, 44*(5), 936–957. http://doi.org/10.1111/j.1467-9248.1996.tb00343.x

Hall, T. & Hubbard, P. (1996). The entrepreneurial city: new urban politics, new urban geographies? *Progress in Human Geography,* 20(2): 153-174.

Hall, T. & Hubbard, P. (1998). *The entrepreneurial city: geographies of politics, regimes and representation.* Manchester: John Wiley and Sons.

Harding, A. (1997). Urban regimes in a Europe of the cities? *European Urban and Regional Studies,* 4, 291-314.

Harris, D. J. (1982). Structural change and economic growth-a review article. *Contributions to Political Economy, 1,* 25–45.

Harvey, D. (1978). The urban process under capitalism: a framework for analysis. *International Journal of Urban and Regional Research, 2*(1-4), 101–131. http://doi.org/10.1111/j.1468-2427.1978.tb00738.x

Harvey, D. (1989). From managerialism to entreprenialism: the transformation in urban governance in late capitalism. *Geografiska Annaler. Series B, Human Geography, 71*(1), 3–17.

Harvey, D. (2001). Spaces of Capital: Towards a Critical Geography. London and New York: Routledge.

Harvey, D. (2004). *Spaces of neoliberalization-towards a theory of uneven geographical development.* (H. Gebhardt & P. Meusburger, Eds.). Wiesbaden: Franz Steiner Verlag.

Harvey, D. (2005). *A Brief History of Neoliberalism.* Oxford, New York: Oxford University Press. http://doi.org/10.1017/CBO9781107415324.004

Harvey, D. (2007a). Neoliberalism and the city. *Studies in Social Justice, 1*(1), 2–13. http://doi.org/papers2://publication/uuid/716879B0-B173-4578-B39C-2A4ACDBBAEC9

Harvey, D. (2007b). Neoliberalism as creative destruction. The American Academy of Political and Social Science, 610(1), 21-44. http://doi.org/10.1177/0002716206296780.

Hasenfeld, Y., & Garrow, E., E. (2012). Nonprofit Human-Service Organizations, Social Rights, and Advocacy in a Neoliberal Welfare State. *Social Service Review*, 86(2), 295-322.

Haughton, G., Allmendinger, P., & Oosterlynck, S. (2013). Spaces of neoliberal experimentation: soft spaces, postpolitics, and neoliberal governmentality. *Environment and Planning A*, *45*(1), 217–234. http://doi.org/10.1068/a45121

He, S., & Chen, G. (2012), Theme issue: Interrogating unequal rights to the Chinese city, *Environment and Planning A,* 44(12). London, UK: Pion

He, S., & Lin, G. C. (2015). Producing and consuming China's new urban space: state, market and society. *Urban Studies*, *52*(15), 2757–2773. http://doi.org/10.1177/0042098015604810

He, S., & Wu, F. (2009). China's emerging neoliberal urbanism: perspectives from urban redevelopment. *Antipode*, *41*(2), 282–304. http://doi.org/10.1111/j.1467-8330.2009.00673.x

He, S., & Wu, F. (2007). Socio-spatial impacts of property-led redevelopment on China's urban neighbourhoods. *Cities*, *24*(3), 194–208. http://doi.org/10.1016/j.cities.2006.12.001

Healey, P. (1997). *Collaborative Planning. Shaping Places in Fragmented Societies*. Houndmills & London: Macmillan Press.

Healey, P. (2004). Creativity and urban governance. *Policy Studies*, *25*(2), 87–102. http://doi.org/10.1080/0144287042000262189

Healey, P. (2006). Transforming governance: challenges of institutional adaptation and a new politics of space. *European Planning Studies*, *14*(3), 299–320. http://doi.org/10.1080/09654310500420792

Heurkens, E. (2012). Private sector-led urban development projects: management, Ppartnerships & effects in the Netherlands and the UK. Create Space Independent Publishing Platform.

Hill, C. J. (2004). Is hierarchical governance in decline? Evidence from empirical research. *Journal of Public Administration Research and Theory*, *15*(2), 173–195. http://doi.org/10.1093/jopart/mui011

Hock, J. (2012). *Political designs: architecture and urban renewal in the civil rights era, 1954-1973*. Harvard University.

Hodgson, G. M. (2006). What are institutions? *Journal of Economic Issues*, *XL*(1), 1–25.

Hoffman, L., DeHart, M., & Collier, S., J. (2006). Notes on the anthropology of neoliberalism. *Anthropol*, 47, 9–10.

Hohn, U., & Neuer, B. (2006). New urban governance: institutional change and consequences for urban development. *European Planning Studies*, *14*(3), 291–298. http://doi.org/10.1080/09654310500420750

Hommels, a. (2000). Obduracy and urban sociotechnical change: changing plan Hoog Catharijne. *Urban Affairs Review*, *35*(5), 649–676. http://doi.org/10.1177/10780870022184589

Hong, S. & Zhang, J. (2012). *Mechanism and governance of urban sprawl-Based on the analysis of economic and institutional*. Nanjing: Southeast University Press, (in Chinese).

Hood, C. (1990). Rediscovering institutions: the organizational basis of political politics by J. G. March; J.P. Olsen. *Journal of Public Policy*, *10*(3), 349–351.

Howard, M. C., & King, J. E. (2008). *The rise of neoliberalism in advanced capitalist economies: a materialist analysis*. New York: Palgrave Macmillan. http://doi.org/10.1017/CBO9781107415324.004

Hsu, J. Y. (2014). Introduction to the special issue on neoliberalism and urban governance in Taiwan. *Journal of Geographical Science*, *72*, 1–3. http://doi.org/10.6161/jgs.2014.72.01

Hudalah, D., Winarso, H., & Woltjer, J. (2016). Gentrifying the peri-urban: land use conflicts and institutional dynamics at the frontier of an Indonesian metropolis. *Urban Studies*, *53*(3), 593–608. http://doi.org/10.1177/0042098014557208

Huntington, S. P., & Liu, N. (1994). Third Wave: the twentieth century wave of democratization. Taipei: *Five Southern Book Publishing Company* (in Chinese).

Hsing, Y. T. (2010). *The great urban transformation: Politics of land and property in China*. Oxford, UK: Oxford University Press.

Hysing, E. (2009). From government to governance? A comparison of environmental governing in swedish forestry and transport.

Governance, *22*(4), 647–672. http://doi.org/10.1111/j.1468-0491.2009.01457.x

Immergut, E. M. (1998). The theoretical core of the new institutionalism. *Politics & Society*, *26*(1), 5–34. http://doi.org/10.1177/0032329298026001002

Jessop, B. (1997). The entrepreneurial city: re-imaging localities, redesigning economic governance, or restructuring capital? In N. Jewson & S. MacGregor (Eds.), *Realising Cities: New Spatial Divisions and Social Transformation* (pp. 28–41). London: Routledge.

Jessop, B. (1998). The narrative of enterprise and the enterprise of narrative: place marketing and the entrepreneurial city. In Hall, T. & Hubbard, P. (Eds.). *The entrepreneurial city: geographies of politics, regimes and representation.* Manchester: John Wiley and Sons, 77–102.

Jessop, B. (2000). Governance failure. In: G. Stoker (Ed.) The New Politics of British Local Governance. London: Macmillan Press, 11–32.

Jessop, B. (2002). Liberalism, neoliberalism, and urban governance: a state-theoretical perspective. *Antipode*, *34*(3), 452–472. http://doi.org/10.1111/1467-8330.00250

Jessop, B. (2001). Institutional (re)turns and the strategic-relation approach. *Environment and Planning A,* 33, 1213–1235.

Jessop, B. (2012). Liberalism, neoliberalism, and urban governance: a state–theoretical perspective. In *Spaces of Neoliberalism* (pp. 104–125). Chichester, UK: John Wiley & Sons, Ltd. http://doi.org/10.1002/9781444397499.ch5

Jessop, B., & Sum, N.-L. (2000). An entrepreneurial city in action: Hong Kong's emerging strategies in and for (inter)urban competition. *Urban Studies*, *37*(12), 2287–2313. http://doi.org/10.1080/00420980020002814

Johnson, L. S. (2005). *Constitutional change in local governance: an exploration of institutional entrepreneurs, procedural safeguards, and selective incentives.* Florida State University.

Jonas K. Ward, A. E. G. (2007). Introduction to a debate on city-regions: new geographies of governance, democracy and social reproduction (debates and developments). *International Journal of Urban and Regional Research, 31*(1), 169–178. http://doi.org/10.1111/j.1468-2427.2007.00711.x

Jonas, A. E. G. (2008). The neoliberal city: governance, ideology, and development in American urbanism, by Jason Hackworth. *Economic Geography, 84*(1), 121–122. Retrieved from http://www.loc.gov/catdir/toc/ecip0617/2006023306.html

Jou, S.-C., Clark, E., & Chen, H.-W. (2016). Gentrification and revanchist urbanism in Taipei? *Urban Studies, 53*(3), 560–576. http://doi.org/10.1177/0042098014541970

Karaman, O. (2010). *Remaking space for globalization: dispossession through urban renewal in Istanbul.* University of Minnesota.

Karaman, O. (2013). Urban renewal in Istanbul: reconfigured spaces, robotic lives. *International Journal of Urban and Regional Research, 37*(2), 715–733. http://doi.org/10.1111/j.1468-2427.2012.01163.x

Kearns, A., & Paddison, R. (2000). New challenges for urban governance. *Urban Studies, 37*(5-6), 845–850. http://doi.org/10.1080/00420980050011118

Kotz, D. M. (2008). *Neoliberalism and financialization.* Retrieved from http://people.umass.edu/dmkotz/Fin_and_NL_08_09.pdf

Kotz, D. M. (2009). The financial and economic crisis of 2008: a systemic crisis of neoliberal capitalism.

Kotz, D. M. (2015). *The Rise and Fall of Neoliberal Capitalism.* Harvard University Press.

Kotz, D. M., & McDonough, T., M.(2010). Global neoliberalism and the contemporary social structure of accumulation. *Contemporary Capitalism and Its Crises: Structure of Accumulation Theory for the Twenty-First Century.* http://doi.org/10.1017/CBO9780511804335.005

Larner, W. (2003). Neoliberalism? *Environment and Planning D: Society and Space, 21*(5), 509–512. http://doi.org/10.1068/d2105ed

Larner, W. (2000). Neo-liberalism: policy, ideology, governmentality. *Studies in Political Economy, 63*, 5–25.

Lauermann, J., & Davidson, M. (2013). Negotiating particularity in neoliberalism studies: tracing development strategies across neoliberal urban governance projects. *Antipode, 45*(5), 1277–1297. http://doi.org/10.1111/anti.12018

Lebaron, F., Campbell, J. L., & Pedersen, O. K. (2002). The rise of neoliberalism and institutional analysis. *Contemporary Sociology, 31*(5), 548. http://doi.org/10.2307/3090039

Lee, J. & Zhu, Y. (2006). Urban governance, neoliberalism and housing reform in China. *The Pacific Review*, 19(1), 39-61.

Lees, L. (2013). The urban injustices of new labour's "new urban renewal": the case of the aylesbury estate in London. *Antipode, 46*(4), 921–947. http://doi.org/10.1111/anti.12020

Lefebvre, H. (1976). The survival of capitalism: reproduction of the relations of production. New York: St. Martin's Press.

Lefebvre, H. (1991). *The production of space. Blackwell*. Oxford, UK & Cambridge, USA: Blackwell.

Lefebvre, H. (2003). The urban revolution. Minneapolis, London: University of Minnesota Press.

Lefebvre, H. (2009). *State, space, world. Critique.* http://doi.org/10.1177/0042098010391297

Leibovitz, Y. (2001). *Associative governance? The political economy of institutional change in two Ontario city regions.* University of Toronto.

Leitner, H. (1990). Cities in pursuit of economic growth: The local state as entrepreneur. *Political Geography Quarterly*, 9 (2), 146-170.

Leitner, H., Peck, J., & Sheppard, E. (2007). *Contesting neoliberalism: urban frontiers. The Guilford Press.* New York, London: The Guilford Press.

Leitner, H., & Sheppard, E. (1998). Economic uncertainty, inter-urban competition and the effeicacy of entrepreneurialism. In Hall, T. & Hubbard, P. (Eds.). *The entrepreneurial city: geographies of politics, regimes and representation.* Manchester: John Wiley and Sons, 285-308.

Lemke, T. (2000). Foucault, governmentality, and critique. In *Rethinking Marxism Conference.* http://doi.org/10.1080/089356902101242288

Lemke, T. (2001). "The birth of bio-politics": Michel Foucault's lecture at the Collège de France on neo-liberal governmentality. *Economy and Society*, *30*(2), 190–207. http://doi.org/10.1080/713766674

Liew, L. (2005). China's engagement with neo-liberalism: path dependency, geography and party self-reinvention. *Journal of Development Studies*, *41*(2), 331–352. http://doi.org/10.1080/0022038042000309278

Lim, K. F. (2013). "Socialism with Chinese characteristics": Uneven development, variegated neoliberalization and the dialectical differentiation of state spatiality. *Progress in Human Geography*, *38*(2), 221–247. http://doi.org/10.1177/0309132513476822

Lin, G. C. S. (2009). Scaling-up regional development in globalizing China: local capital accumulation, land-centred politics, and reproduction of space. *Regional Studies*, *43*(3), 429–447. http://doi.org/10.1080/00343400802662625

Lin, G.C.S. (2014). China's landed urbanization: neoliberalizing politics, land commodification, and municipal finance in the growth of metropolises. *Environment and Planning A, 2014(46)*, 1814-1835. http://doi.org/10.1068/a130016p

Lin, Y., Pu, H., & Geertman., S. (2015). A conceptual framework on modes of governance for the regeneration of Chinese 'villages in the city'. *Urban Studies*, 52 (10), 1774-1790.

Lin, G. C. S., & Ho, Samuel, P. S. (2005). The state, land system, and land development processes in contemporary China. *Annals of the Association of American Geographers*, *95*(2), 411–436. http://doi.org/10.1111/j.1467-8306.2005.00467.x

Lin, G. C.S., & Zhang, a. Y. (2015). Emerging spaces of neoliberal urbanism in China: Land commodification, municipal finance and local economic growth in prefecture-level cities. *Urban Studies*, *52*(15), 2774–2798. http://doi.org/10.1177/0042098014528549

Logan, J. R., & Molotch, H. L. (1987). *Urban fortunes: the political economy of place*. Berkeley: University of California Press.

Lowndes, V. (2001). Rescuing Aunt Sally: taking institutional theory seriously in urban politics. *Urban Studies*, *38*(11), 1953–1971. http://doi.org/10.1080/00420980120080871

Luo, Y. (2012). A restricted view of contemporary China's neoliberalism. *Modern Philosophy*, 2012(5), 60-65 (in Chinese).

Macleavy, J. (2012). The lore of the jungle: neoliberalism and statecraft in the global-local disorder (revisiting Peck and Tickell). *Area*, *44*(2), 250–253. http://doi.org/10.1111/j.1475-4762.2012.01093.x

MacLeod, G. (2001). New regionalism reconsidered: globalization and the remaking of economic space. *International Journal of Urban and Regional Research*, 25(4), 804-829

MacLeod, G., & Goodwin, M. (1999). Reconstructing an urban and regional political economy: on the state, politics, scale, and explanation. *Political Geography*, 18, 697–730.

MacLeod, G., & Jones, M. (1999). Re-regulating a regional rustbelt: institutional fixes, entrepreneurial discourse, and the "politics of representation". *Environment and Planning D*: Society and Space, 17, 575-605.

Marx, K. (1887). Capital : a critique of political economy Volume I Book One: the process of production of capital. *Kapital English*, *I* (2008), 543. http://doi.org/10.1002/ejoc.201200111

Mayer, M., & Künkel, J. (2012): *Neoliberal Urbanism and its Contestations: Crossing Theoretical Boundaries.* London, UK: Palgrave Macmillan.

McCann, E. (2004). "Best places": interurban competition, quality of life and popular media discourse. *Urban Studies*, *41*(10), 1909–1929. http://doi.org/10.1080/0042098042000256314

McCann, E., & Ward, K. (2011), *Mobile Urbanism: Cities and Policymaking in the Global Age*. Minneapolis: University of Minnesota Press.

McCarthy, J., & Prudham, S. (2004). Neoliberal nature and the nature of neoliberalism. *Geoforum*, *35*(3), 275–283. http://doi.org/10.1016/j.geoforum.2003.07.003

Mcclure, L., Baker, D., & Faculty, E. (2013). Doing adaptation differently? Does neoliberalism influence adaptation planning in Queensland? *State of Australian Cities Conference*, 1–9.

McGuirk, P. M. (2000). Power and policy networks in urban governance: local government and property-led regeneration in Dublin. *Urban Stud*, *37*(4), 651–672. http://doi.org/10.1080/00420980050003955

Menger, C. (1983). *Investigations into the Method of the Social Sciences, with Special Reference to Economics, New York and London*: New York University Press.

Monstadt, J. (2007). Urban governance and the transition of energy systems: institutional change and shifting energy and climate policies in Berlin. *International Journal of Urban and Regional Research*, *31*(2), 326–343. http://doi.org/10.1111/j.1468-2427.2007.00725.x

Morange, M. (2011). Right to the city, neoliberalism and the developmental state in Cape Town. *Justice Spatiale Spatial Justice*, 1–15.

Miller, P., & Rose, N. (1990). The government of economic life. *Economy and Society*, 19, 1-31.

Miro, S. V. (2011). Producing a "Successful City": Neoliberal Urbanism and Gentrification in the Tourist City—The Case of Palma. *Urban Studies Research*, 2011, 1-13.

Mudge, S. (2008). What is neoliberalism?. *Socio-economic Review*, 6, 703-731.

North, D. C. (1990). *Institutions, institutional change, and economic performance*. Cambridge, UK: Cambridge University Press.

North, D. C. (1991). Institutions. *Journal of Economic Perspectives*, *5*(1), 97–112.

Oi, J. C. (1995). The role of the local state in China's transitional economy. *China Quarterly*, 144,1132–49.

Olsen, J. P. (2008). The Ups and Downs of Bureaucratic Organization. *Annual Review of Political Science*, *11*(1), 13–37. http://doi.org/10.1146/annurev.polisci.11.060106.101806

Olson, M. (1965). *The logic of collective action*. London, UK: Harvard University Press.

Ong, A. (2006). *Neoliberalism as Exception: Mutations in Citizenship and Sovereignty*, Duke University Press.

Ong, A. (2007). Neoliberalism as a mobile technology. *Transactions of the Institute of British Geographers*, 32(1), 1-8.

Ong, A.(2007). Boundary crossings: neoliberalism as a mobile technology. *Trans Inst Br Geogr*, *32*, 3–8. http://doi.org/10.1111/j.1475-5661.2007.00234.x

Ostrom, E. (2005). *Understanding institutional diversity*. Princeton University Press, Princeton, NJ

Paddison, R. (1993). City marketing, image reconstruction and urban regeneration. *Urban Studies*, 30(2), 339-350.

Parker, J. (2012). Unravelling the neoliberal paradox with Marx. *Journal of Australian Political Economy*, 70, 193–213.

Parkinson, M., Foley, B., and Judd, D. (1988). *Regenetating the Cities: The U.K. Crisis and the U.S. Experience*. Manchester: Manchester University Press.

Peck, J., & Tickell, A. (1995). Business goes local: dissecting the 'business agenda' in Manchester. *International Journal of Urban and Regional Research*,19 (1), 55-78.

Peck, J. (2004). Geography and public policy: constructions of neoliberalism. *Progress in Human Geography*, *28*(3), 392–405. http://doi.org/10.1191/0309132504ph492pr

Peck, J. (2008). Remaking laissez-faire. *Progress in Human Geography*, *32*(1), 3–43. http://doi.org/10.1177/0309132507084816

Peck, J. (2013). Excavating the pilbara: a Polanyian exploration. *Geographical Research*, *51*(3), 227–242. http://doi.org/10.1111/1745-5871.12027

Peck, J., Theodore, N., & Brenner, N. (2009). Neoliberal urbanism: models, moments, mutations. *SAIS Review*, *29*(1), 49–66. http://doi.org/10.1353/sais.0.0028

Peck, J., Throdore, N., & Bernner, N. (2013). Neoliberal urbanism redux? International Journal of Urban and Regional Research, 37(3), 1091-1099.http://doi.org/10.1111/1468-2427.12066

Peck, J., & Tickell, A. (2002). Neoliberalizing space. *Antipode*, *34*(3), 380–404. http://doi.org/10.1111/1467-8330.00247

Peck, J., & Zhang, J. (2013). A variety of capitalism ... with Chinese characteristics? *Journal of Economic Geography*, *13*(3), 357–396. http://doi.org/10.1093/jeg/lbs058

Peterson, P. E. (1981). *City limits*. Chicago : University of Chicago Press.Perez, P. C. R. (2015). *Nature under capitalism: Neil Smith's proposal of the production of nature*. Retrieved from http://espaiscritics.org/wp-content/uploads/2015/09/CEPERO.pdf

Picton, R. M. (2009). *"A capital experience": national urban renewal, neoliberalism, and urban governance on LeBreton Flats in Ottawa, Ontario, Canada*. University of Toronto.

Pierre, J. (1999). Models of urban governance: the institutional dimension of urban politics. *Urban Affairs Review*, *34*(3), 372–396. http://doi.org/10.1177/10780879922183988

Peter, B. G., & Pierre, I. (2006). Governnance, government and the state. In the State, Hay, C., Lister, M., & Marsh. D. Basinstoke: Palgrave, 209-222.

Polanyi, M. (1967). *The Tacit Dimension*. London: Routledge and Kegan Paul.

Qian, Y., & Weingast., B. R. (1997). Federalism as a commitment to preserving market incentives. *Journal of Economic Perspectives*, 11 (4), 83–92.

Raco, M., & Imrie, R. (2000). Governmentality and rights and responsibilities in urban policy. *Environment and Planning A*, 32(12), 2187–2204. http://doi.org/10.1068/a3365

Ramstrom, J., Jansson, H., & Johanson, M. (2006). Institutions and networks: business network logic in the Chinese, Russian and West European institutional contexts. *In 22nd Industrial Marketing and Purchasing Group Conference*. Retrieved from http://www.impgroup.org/uploads/papers/5740.pdf

Ramsay, M. (1996). The local community-market of culture and wealth. *Journal of Urban Affairs*, *18*(2), 95–118.

Richter, R. (2015). *Essays on new institutional economics*. Cham: Springer International Publishing. http://doi.org/10.1007/978-3-319-14154-1

Robinson, J. (2011). The travels of urban neoliberalism: Taking stock of the internationalization of urban theory. *Urban Geography*, 32 (8): 1087–1109

Robinson, J., & Roy, A. (2015). Global urbanisms and the nature of urban theory. *International Journal of Urban and Regional Research*. http://doi.org/10.1111/1468-2427.12272

Ross, M. H. (1997). Culture and identity in comparative political analysis. In M. I. Lichbach & A. S. Zuckerman, Comparative politics: Rationality, culture, and structure. Cambridge, UK: Cambridge Univ. Press, 42-80.

Rose, N., & Miller, P. (1992). Political power beyond the State: problematics of government. *British Journal of Sociology*, 43, 173-205.

Roy, A., & Ong, A. (2011). *Worlding Cities: Asian Experiments and the Art of Being Global*. Oxford : Wiley-Blackwell.

Savitch, H. V. (1998). Global challenge and institutional capacity: or, how we can refit local administration for the next century. *Administration & Society*, *30*(3), 248–273. http://doi.org/10.1177/0095399798303002

Scharpf, F. W. (1989). Decision rules, decision styles and policy choices. *Journal of Theoretical Politics*, *1*(2), 149 –176. http://doi.org/10.1177/0951692889001002003

Schotter, A. (1981). Why take a game theoretical approach to economics? Institutions, economics and game theory. *The C.V. Starr Center for Applied Economics*. Retrieved from https://www.researchgate.net/publication/4857881_Why_Take_a_G ame_Theoretical_Approach_to_Economics_Institutions_Economics _and_Game_Theory

Schotter, A., & Schwödiauer, G. (1980). Economics and the theory of games: a survey. *Journal of Economic Literature*, *18*(2), 479–527. http://doi.org/10.2277/0521772516

Schumpeter, J.A. (1934), *The Theory of Economic Development: An Inquiry into Profits, Capital, Credit, Interest, and the Business Cycle*. London: Transaction Publishers.

Selznick, P. (1996). Institutionalism "old" and "new." *Administrative Science Quarterly, 41*(2), 270–277.

Schimank, U. (2004). Handeln in Institutionen und handelnde Institutionen, in: F. Jaeger & J. Straub (Eds) Handbuch der Kulturwissenschaften, 2: Paradigmen und Disziplinen. Stuttgart: Weimar:J.B. Metzler, 293–307.

Sheppard, E. (2015). Thinking geographically: globalizing capitalism and beyond. *Annals of the Association of American Geographers, 105*(6), 1113–1134. http://doi.org/10.1080/00045608.2015.1064513

Shepsle, K. A. (1986). Institutional equilibrium and equilibrium institutions. In H. F. Weisberg (Ed.), *Political Science: The Politics of Science* (pp. 51–81). New York, USA: Agathon Press, Inc.

Shin, H. B. (2014). Contesting speculative urbanisation and strategising discontents. *City, 18*(4-5), 509–516. http://doi.org/10.1080/13604813.2014.939471

Shin, H. B. (2015). Economic transition and speculative urbanisation in China: gentrification versus dispossession. *Urban Studies, 53*(3), 471–489. http://doi.org/10.1177/0042098015597111

Short, J.R., & Kim, Y-H. (1998). Urban crises/urban representations: selling the city in difficult times. In Hall, T. & Hubbard, P. (Eds.). *The entrepreneurial city: geographies of politics, regimes and representation.* Manchester: John Wiley and Sons, 55-76.

Shirk, S. L. (2007). China fragile superpower. Oxford: Oxford University Press.

Smith, N. (1990). The production of nature. In Smith, N. (Eds.), *Uneven Development: Nature, Capital and the Production of Space.* Oxford, UK: Blackwell, 34–65.

Smith, N. (2002). New globalism, new urbanism: gentrification as global urban strategy. *Antipode, 34*(3), 427–450. http://doi.org/10.1111/1467-8330.00249

Smith, N. (2010). *Uneven development: Nature, capital, and the production of space.* Oxford, UK: Basil Blackwell.

Soja, E. (1989). Postmodern geographies. *The Reassertion of Space in Critical Social Theory.* London & New York: Verso. http://doi.org/910.4 SOJ

Soja, E. (1996). *Thirds pace: journeys to Los Angeles and other real-and-imagined places*. Cambridge, Mass.: Blackwell.

Song, W., & Zhu, X. (2010). Gentrification in urban China under market transformation. *International Journal of Urban Sciences*, *14*(2), 152–163. http://doi.org/10.1080/12265934.2010.9693673

Sotiropoulos, D. P. (2011). Kalecki's Dilemma: Toward a Marxian Political Economy of Neoliberalism. *Rethinking Marxism: A Journal of Economics, Culture & Society*, 23(1), 100-116.

Steinmo, S. (2011). Institutionalism. In Polsby, N. (Eds.), *International Encyclopedia of the Social and Behavioral Sciences*. Elsevier Science.

Storper, M., & Scott, A. J. (2016). Current debates in urban theory: a critical assessment. *Urban Studies*, 1–36. http://doi.org/10.1177/0042098016634002

Su, X. (2015) Urban Entrepreneurialism and the Commodification of Heritage in China, *Urban Studies*, 52, 2874-2889.

Swyngedouw, E. (2004). Globalisation or "glocalisation"? Networks, territories and rescaling. *Cambridge Review of International Affairs*, *17*(1), 25–48. http://doi.org/10.1080/0955757042000203632

Swyngedouw, E. (2005). Governance innovation and the citizen: the Janus face of governance-beyond-the-state. *Urban Studies*, *42*(11), 1991–2006. http://doi.org/10.1080/00420980500279869

Swyngedouw, E., Moulaert, F., & Rodriguez, A. (2002). Neoliberal urbanization in Europe: large–scale urban development projects and the new urban policy. *Antipode*, *34*(3), 542–577. http://doi.org/10.1111/1467-8330.00254

Tasan-Kok, T., & Zaleczna, M. (2010). *Public-Private Partnerships in urban development projects. Polish practices and EU regulations*. Warsaw: Ernst & Young Poland.

Thiagarajan, M., Coote-solek, E. W., Mireya, B., New, N. N., Times, Y., Historical, P., … Reproduced. (2007). *Reproduced with permission of the copyright owner. Further reproduction prohibited without permission. New York*. http://doi.org/10.3102/00346543067001043

Thorsen, D. E. (2009). *The neoliberal challenge: what is neoliberalism?* Retrieved from http://ctx.sagepub.com/lookup/doi/10.1525/ctx.2007.6.3.13

Thorsen, D., E., & Lie, A. (2007): "Kva er nyliberalisme?". In D. H. Claes et al. (Eds.): Nyliberalisme – ideer og politisk virkelighet. Oslo: Universitetsforlaget, 33-48.

Tochterman, B. (2012). Theorizing neoliberal urban development: a genealogy from Richard Florida to Jane Jacobs. *Radical History Review*, (112), 65–87. http://doi.org/10.1215/01636545-1416169

Uitermark, J. (2005). The genesis and evolution of urban policy: a confrontation of regulationist and governmentality approaches. *Political Geography*, *24*(2), 137–163. http://doi.org/10.1016/j.polgeo.2004.07.009

Veblen, T. (1899). T*he Theory of the Leisure Class: An Economic Study of Institutions*, New York: Macmillan.

Vogel, E. (2011), *Deng Xiaoping and the transformation of China*. Hravard: Belknap Press.

Miró, S. V. (2011). Producing a "successful city": neoliberal urbanism and gentrification in the tourist city-the case of Palma (Majorca). *Urban Studies Research*, 1–13. http://doi.org/10.1155/2011/989676

Wai Wong, S. (2015). Urbanization as a process of state building: local governance reforms in China. *International Journal of Urban and Regional Research*, 1–15. http://doi.org/10.1111/1468-2427.12250

Walder, A. G. (1995). Local governments as industrial firms: an organizational analysis of China's transitional economy. *American Journal of Sociology*, *101*(2), 263–301.

Waley, P. (2015). Speaking gentrification in the languages of the Global East. *Urban Studies*, *53*(3), 615–625. http://doi.org/10.1177/0042098015615726

Walliser, B. (2006). Game theory and emergence of institutions. *Mimeo*. Retrieved from http://www.pse.ens.fr/users/walliser/pdf/games.pdf

Walliser, 2006; Etzold, B., Jülich, S., Keck, M., Sakdapolrak, P., Schmitt, T. und Zimmer, A. (2012): *Doing institutions. New trends in institutional theory and their relevance for development geography. In: Erdkunde, 66 (3), 185-195.*

Wang, C. (2009). Dialectics in multitude an exploration into/beyond Henri Lefebvre's conceptual triad of production of space. *Journal of Geographical Science*, *24*, 1–24.

Wang, H. (2008). *The rise of contemporary Chinese thought.* Beijing: SDX Joint Publishing (in Chinese).

Wang. T., & Li, B. (2008). The state ideal of egalitarianism and the bureaucratic system of articularism: a new model of social stratification structure. *Chinese Journal of Society*, 2008(5), (in Chinese).

Wang, Y., & Murie. A. (1996). The process of commercialisation of urban housing in China. *Urban Studies*, 33 (6), 971-989.

Weber, R. (2002). Extracting value from the city: neoliberalism and urban redevelopment. *Antipode*, *34*(3), 519–540. http://doi.org/10.1111/1467-8330.00253

Wei, K. (1991). *Lanzhou Economic History.* Lanzhou: Lanzhou University Press (in Chinese).

Wei, Y. H. D. (2012). Restructuring for growth in urban China: transitional institutions, urban development, and spatial transformation. *Habitat International*, *36*(3), 396–405. http://doi.org/10.1016/j.habitatint.2011.12.023

Weller, S., & O'Neill, P. (2014). An argument with neoliberalism: australia's place in a global imaginary. *Dialogues in Human Geography*, *4*(2), 105–130. http://doi.org/10.1177/2043820614536334

Wilson, D. (2004). Toward a contingent urban neoliberalism. *Urban Geography*, *25*(8), 771–783. http://doi.org/10.2747/0272-3638.25.8.771

Wong, C. P. W. (2000). Central-local relations revisited: the 1994 tax-sharing reform and public expenditure management in China. *China Perspectives*, 52–63.

Wright, I. (2013). Are we all neoliberals now? Urban planning in a neoliberal era. In *49th ISOCARP Congress 2013.*

Wu, F. (2002). China's changing urban governance in the transition towards a more market-oriented economy. *Urban Studies, 39*(7), 1071–1093. http://doi.org/10.1080/0042098022013549

Wu, F. (2003). The post-socialist entrepreneurial city as the state project: Shanghai's globalisation in question. *Urban Studies, 40*(9), 1673–1698. http://doi.org/10.1080/0042098032000106555

Wu, F. (2008). China's great transformation: neoliberalization as establishing a market society. *Geoforum, 39*(3), 1093–1096. http://doi.org/10.1016/j.geoforum.2008.01.007

Wu, F. (2010). How neoliberal is China's reform? The origins of change during transition. *Eurasian Geography and Economics, 51*(5), 619–631. http://doi.org/10.2747/1539-7216.51.5.619

Wu, F. (2016). Emerging Chinese cities: implications for global urban studies, *The Professional Geographer*, 68:2, 338-348, DOI: 10.1080/00330124.2015.1099189

Wu, F., Xu, J., & Yeh, A.G.O. (2007). *Urban redevelopment in post-reform China: state. Market, and space.* London and New York : Routledge.

Xie, Z. (1999).*To break the siege: China housing reform action.* Beijing: Social Sciences Academic Press (in Chinese).

Xu, J., & Yeh, A. G. O. (2005). City repositioning and competitiveness building in regional development: new development strategies in Guangzhou, China. *International Journal of Urban and Regional Research, 29*(2), 283–308. http://doi.org/10.1111/j.1468-2427.2005.00585.x

Yang, L., & Wang, Y. (1992). *Housing reform: theoretical reflection and realistic option.* Tianjin: Tianjin People's Publishing House (in Chinese).

Yang, Y., & Yang, X. (2009). Research on urban spatial expansion and land-use inner structure transformation of the Large Valley-basin Cities in China from 1949 to 2005-A Case Study of Lanzhou. *Journal of natural resources*, 24(1), 37-49, (in Chinese).

Yeh, A. G. O. & Wu. F (1996). The new land development process and urban development in Chinese cities. *International Journal of Urban and Regional Research*, 20 (2), 330-353.

Ye, L., & Wu, A. M. (2014). Urbanization, land development, and land financing: evidence from Chinese cities. *Journal of Urban Affairs*, *36*(S1), 354–368. http://doi.org/10.1111/juaf.12105

Yong, W. (2003). *An institutional analysis of China's urban development: case study of real estate market in transitional Chongqing. National University of Singapore.* National University of Singapore.

Yu, B., & Zhen, X. (2003). Arable land crisis and blindness in the urbanization processes. *Urban Issues*, 113(3), 58-62, (in Chinese).

Yuan, H. (2006). Debate over neo-liberalism and the New Left: the essence of contemporary Chinese political reformism. *Huang Hua Gang Magazine*. 1(16), 1-4 (in Chinese).

Zhang, X. (1997). Urban land reform in China. *Land Use Policy*, 14 (3), 187-199.

Zhang, J. (2012). From Hong Kong's capitalist fundamentals to Singapore's authoritarian governance: the policy mobility of neo-liberalising Shenzhen, China. *Urban Studies*, *49*(13), 2853–2871. http://doi.org/10.1177/0042098012452455

Zhang, J. (2013). Marketization beyond neoliberalization: A neo-Polanyian perspective on China's transition to a market economy. *Environment and Planning A*, *45*(7), 1605–1624. http://doi.org/10.1068/a45589

Zhao, Y. & Bourassa, S. (2003). China's urban housing reform: recent chievements and new inequalities. *Housing Studies*, 18(5), 721-744.

Zhu, J. (1999). Local growth coalition: the context and implications of China's gradualist urban land reforms. *International Journal of Urban and Regional Research*, *23*(3), 534–548. http://doi.org/10.1111/1468-2427.00211

Zhu, J. (2005). A transitional institution for the emerging land market in urban China. *Urban studies*, 42(8), 1369-1390.

Zhu, P. (2015). Residential segregation and employment outcomes of rural migrant workers in China. *Urban Studies*. http://doi.org/10.1177/004209801557861

Geographie: Forschung und Wissenschaft

Hans Gebhardt (Ed.)
Urban Governance, Spatial Planning and Economic Development in the 21th Century China
China's cities are subject to dramatic changes. Cities develop into Megacities, economic growth as well as the drastic increase of traffic contribute to a profound transformation of urban infrastructure. However, the processes are more visible than the stakeholders supporting such transformations. What are the location factors, spatial principles and planning philosophies that direct the cities' growth and reconstruction?
The articles of this anthology investigate the above mentioned questions. Using various case studies, they analyse processes of location choice and transformation in Chinese coastal Megacities and in inland areas; they explore urban governance processes and - vice versa - also include the planning concepts of rural areas.
Bd. 6, 2018, 224 S., 44,90 €, br., ISBN 978-3-643-90418-8

Berliner China-Hefte
Chinese History and Society
Edited by Mechthild Leutner (FU Berlin)

Jens Damm; Hauke Neddermann (Eds.)
Intercultural Dialogue across Borders
China between Tradition and Modernity
This issue of the Berliner Chinahefte/Chinese History and Society deals with cultural exchanges between China and the outside world and with their impact, mostly in terms of questions regarding tradition and modernity.
China is understood more as an area composed of certain cultural elements which may include the Chinese (and Taiwanese) diaspora, where a sense of belonging still exists, and which exerts influence on everyday culture and habits.
Bd. 51, 2020, 178 S., 29,90 €, br., ISBN 978-3-643-91254-1

Jens Damm; Mechthild Leutner; Hauke Neddermann (Eds.)
China in a Global Context
Perspectives on and from China
China as a place of isolation hiding from the world behind a great wall – an image which has become exposed as a pejorative cliché. However, The opposite is the case: The country has been immersed in transnational and transcultural exchange for centuries. China was, is and will be part of a global context. This volume sheds new light on the dimensions of China's continuous engagement with the world: as an inspiration for European garden designers, as a participant in global dialogue on culture, law and innovative technology, as a destination for academics, business leaders and refugees, but also – and increasingly so – as a powerful political actor in Asia and throughout the world.
Bd. 50, 2018, 166 S., 29,90 €, br., ISBN 978-3-643-91065-3

Jens Damm; Mechthild Leutner; Niu Dayong (Eds.)
China's Interaction with the World
Historical and Contemporary Aspects
The rapidly changing role of China – once an isolated pariah state, now a G-20 member and an emerging superpower in Asia and beyond – is one of the factors to be considered in any conceptualization of the current state of global affairs. The articles in this issue offer preliminary insights into the expansive topic of China's diversified economic, political and cultural interactions with the world. U.S. policies towards Tibet during the Cold War period are examined as well as current global Chinese business networks, China's foreign policy in the 21st century, and the developing relations between China and the five Central Asian states.
Bd. 49, 2017, 160 S., 29,90 €, br., ISBN 978-3-643-90960-2

Clemens von Haselberg; Stefan Kramer (Hg.)
Zeit, Raum und die Wirklichkeiten Chinas
Bd. 48, 2017, 196 S., 29,90 €, br., ISBN 978-3-643-13620-6

LIT Verlag Berlin – Münster – Wien – Zürich – London
Auslieferung Deutschland / Österreich / Schweiz: siehe Impressumsseite

Berliner China-Studien

hrsg. von Prof. Dr. Mechthild Leutner (Freie Universität Berlin)

Xue Li
Making Local China
A Case Study of Yangzhou, 1853 – 1928
This book presents a case study on the city of Yangzhou in China from 1853 to 1928. During this time, the local society of Yangzhou experienced profound changes towards modernization, when the nation-state of China gradually took shape at the local level. "Yangzhou under the Qing" was transformed into "Yangzhou under modern China". The diverse interactions between the Protestant missions and the multiple actors in the local society kept generating new local context and giving special input to the shaping of modernity in the local society. This study analyses the changing situations of the local society as well as the role of Protestant Church as part of the local social fabric, and tries to achieve a better understanding of how modern China developed out of armed conflicts, power-play, and cooperations among different actors in the local society.
Bd. 56, 2018, 322 S., 39,90 €, br., ISBN 978-3-643-90894-0

Mechthild Leutner
Kolonialpolitik und Wissensproduktion
Carl Arendt (1838 – 1902) und die Entwicklung der Chinawissenschaft
Bd. 55, 2016, 736 S., 69,90 €, br., ISBN 978-3-643-13592-6

Ryanne Flock
Shikumen Linong
Wohnraum und urbaner Wandel im modernen Shanghai
Bd. 54, 2020, 256 S., 29,90 €, br., ISBN 978-3-643-13023-5

Mechthild Leutner; Andreas Steen; Xu Kai; Xu Jian; Jürgen Kloosterhuis; Hu Wanglin; Hu Zhongliang (Hg.)
Preußen, Deutschland und China
Entwicklungslinien und Akteure (1842 – 1911). Redaktion: Yang Zhan'ao
Bd. 53, 2014, 384 S., 39,90 €, br., ISBN 978-3-643-12487-6

Christoph Grau
Die Frühphase der Psychiatrie Chinas
Vergleichende Betrachtung der ersten psychiatrischen Anstalten im China der späten Kaiser- und Republikzeit (1898 – 1945)
Bd. 52, 2014, 152 S., 24,90 €, br., ISBN 978-3-643-12478-4

Sinologie

Xiaoyan Liu
The Changing Face of Women's Education in China
Telling the Story of St. Mary's Hall, McTyeire School and Shanghai No. 3 Girls' Middle School
This book offers a critical study on the history of Shanghai No.3 Girls' Middle School, from its missionary predecessors, St. Mary's Hall and McTyeire School, to its present form as a public school. By bringing together three historical periods, late imperial, the Republic of China and the People's Republic of China, and their respective political regimes into one project and tracing continuities and discontinuities in terms of education between the Nationalists and Communists, the book argues that education in Chinese modern history affords another example of "continuous revolution".
Bd. 5, 2017, 366 S., 39,90 €, br., ISBN 978-3-643-90817-9

LIT Verlag Berlin – Münster – Wien – Zürich – London
Auslieferung Deutschland / Österreich / Schweiz: siehe Impressumsseite